Hidden in the Heavens

Hidden in the Heavens

HOW THE KEPLER MISSION'S QUEST
FOR NEW PLANETS CHANGED
HOW WE VIEW OUR OWN

JASON STEFFEN

PRINCETON UNIVERSITY PRESS

PRINCETON & OXFORD

Requests for permission to reproduce material from this work should be sent to permissions@press.princeton.edu

Published by Princeton University Press
41 William Street, Princeton, New Jersey 08540
99 Banbury Road, Oxford OX2 6JX

press.princeton.edu

ISBN 9780691242477
ISBN (e-book) 9780691242484

British Library Cataloging-in-Publication Data is available

Editorial: Abigail Johnson
Production Editorial: Mark Bellis
Jacket Design: Karl Spurzem
Production: Danielle Amatucci
Publicity: Matthew Taylor and Kate Farquhar-Thomson

Jacket Images: Courtesy of NASA

This book has been composed in Arno

Printed on acid-free paper. ∞

Printed in the United States of America

10 9 8 7 6 5 4 3 2 1

To William Borucki and David Koch for their tenacious work in bringing the Kepler mission to life.

It was my pleasure to work with all of the members of the broader Kepler mission, from the personnel at NASA, to the engineers at Ball, to the scientists on the science team, and the students and postdocs who worked with us. The full scope of the scientific importance of this mission, and the personal contributions of all of those who were part of the mission, cannot be expressed in the limited pages of this text. Suffice it to say that being part of Kepler was a remarkable experience, and I am grateful for having had the opportunity. I especially thank those members of the science team who shared their thoughts with me over our many years of collaboration and more recently as I prepared this work.

CONTENTS

1

Introduction

On a late evening in March 2009, a group of people stood near the beach on the Atlantic coast of Florida. As they exchanged smiles, they nervously monitored the skies to the south. In the distance, a new space telescope was perched atop a large rocket set to lift off. This long-awaited launch was the culmination of twenty-five years of work from an ever-expanding group of scientists and engineers, some of whom, including me, were standing on that beach looking toward the launchpad. Our mission would help answer some of the most profound questions that humanity has ever posed: Are there other worlds like ours out in the emptiness of space? Are we alone?

This craft was slated to make its observations for three and a half years, though the more optimistic among the group thought it might fly for nearly a decade. Designed to detect the presence of planets orbiting distant Sun-like stars, the hope was to measure the number of Earth-like planets circulating in Earth-like orbits within our galaxy. If the mission functioned roughly as expected, it would fundamentally change how we viewed our home planet, and ourselves.

The distant flash and rising column of light and smoke marked the transition from concept to reality for this mission. Those who were gathered to see its launch knew that these were the frightening few moments where things could go terribly wrong. As the seconds ticked by, the milestones for the launch came and went with the rocket and its payload performing "nominally"—a good sign. Cheers, hugs, and slapped backs grew more frequent as the rocket flew south over the Caribbean and disappeared. Now, there was nothing left for us to do but return to our hotel rooms and wait to hear the fate of the spacecraft. It still had miles to go before it was safely in its orbit and NASA's Kepler mission could start taking data.

Eight years earlier, in 2001, I set foot on campus at the University of Washington in Seattle as a new graduate student in the physics department. I had turned down offers from several famous schools back east (including my dream school) to attend the "U-Dub." After visiting the other potential graduate programs, and after a lot of discussion, my wife and I felt that Seattle was the place for us. The campus and city were familiar sites since, a few summers prior, I interned at the University of Washington in the Institute for Nuclear Theory, where I worked on solar neutrinos. This was one of those big decisions couples often make before their first anniversary.

We spent the months prior to our move working out the details of my graduate education—choosing what classes I would take, where we would live, and with whom I would work on my research. The plan was to study cosmology. It sounded pretty cool, and I was told that there was a world-renowned cosmologist in the department. For the next half decade, I would work with him, doing whatever it was cosmologists did. Unbeknownst to me, this plan was foiled long before I trekked north from Salt Lake City. The world-renowned cosmologist was promoted to the dean of the college and stopped taking new students. With my bags still packed from the move, my imagined advisor and research area were off the table and I needed to find an alternative advisor and an alternative area of study.

During my first year in Seattle, I tagged along with a research group that studied the properties of distant galaxies, looking at the kinds of stars that they held and how those stars moved about. Then, for another year, I worked with a physics research group, building experiments to test alternative theories of gravity using a glass *torsion pendulum*. While these topics were interesting, and the professors were smart and engaging, neither project quite matched the romantic view that I had rattling around my head for what I would study, and where I would make my contribution to the volume of human knowledge.

Eventually, one professor suggested I look into the new guy, Eric Agol, who had just been hired by the Astronomy Department. He was filling the spot vacated by the recently promoted dean. His work sounded sufficiently cosmology-ish. So I reached out, we exchanged a couple of emails, and, sight unseen, we agreed to work together upon his arrival in the fall of 2003.

From our emails, we already knew what my dissertation project would be, and it was awesome. We were going to make pictures of the supermassive black holes that are found deep in the centers of galaxies. At the time, there was talk in bigwig astronomy circles about imaging those black holes using a network of radio telescopes that spanned the globe. With our project, we would be poised to produce the first-ever image of one—an incredible thought to consider as a graduate student.

While I was thrilled with the plan, there was one small issue. I hadn't yet taken general relativity, a class about Einstein's theory of gravity, which would provide essential background information for researching black holes and their environs. So, to pass the time, my advisor outlined a different problem, a "practice problem." It was a scenario he had discussed with a friend, relating to a recent discovery from a different astronomical discipline. Apparently, a couple of years prior (in the year 2000), astronomers had been observing the star HD 209458 and saw the signal of an orbiting planet as it transited, or passed in front of the star. My advisor thought there might be interesting things we could learn about such planetary systems by looking at small changes in the motions of the planets. It was the first time such an event had been seen, and there were lots of unexplored questions that transiting planets might answer.

Indeed, this entire area of astronomy was new to the scene. In late 1995, when *Toy Story* was showing on the big screen and "Gangsta's Paradise" first hit the airwaves, there were only nine known planets— Mercury, Venus, Earth, Mars, Jupiter, Saturn, Uranus, Neptune, and 51 Pegasi b. The last planet in that list, 51 Pegasi b (or "51-Peg"), was the first planet discovered outside our solar system that orbited a star like the Sun. A pair of Swiss astronomers found it circling a star in the constellation of Pegasus some fifty light-years (three hundred trillion miles) away [1]. The planet had a mass roughly half that of Jupiter, or a bit more than Saturn, and the star it orbited was similar to the Sun in size, mass, and temperature.

While the discovery of 51-Peg was clearly groundbreaking, most astronomers believed that with the right instrument and the right set of observations, the discovery of a planet orbiting a distant star, or an *exoplanet*, was just a matter of time. However, the astronomy community was still taken aback by the 51-Peg discovery, not because they didn't expect there to be other planets, and not because its discovery wasn't a major breakthrough, but because they didn't expect the planet to look like what was found. It was like some inexplicable creature had emerged from the swamp and knocked on our door.

The presumption was that a planet with a mass in the range of Jupiter and Saturn would be similar in structure and composition to these gigantic bodies—mostly gaseous material with a rocky or metallic core. Unlike Jupiter and Saturn, which respectively take twelve and twenty-nine years, to orbit the Sun, 51-Peg orbits its host star every four days. In the solar system, no planet orbits that close to the Sun, and those that come nearest are dense, rocky planets with thin atmospheres. This planet circles its star at one-tenth the distance of Mercury's orbit—along with its huge atmosphere of volatile material that shouldn't have been able to condense under the intense heat.

This discovery upended our astronomical logic. The territory for such giant planets should be the outer parts of a planetary system, where the temperatures are cold. For instance, Jupiter, the closest to the Sun of the solar-system giant planets, is five times farther from the Sun than the Earth. It receives only four percent of the light that we receive. This relatively cold environment allowed the forming Jupiter's gravity to trap lighter elements in its atmosphere. But 51-Peg orbits in less than a week. The blazing temperatures that are found that close to the star should have prevented it from forming. A *hot Jupiter* like 51-Peg ought not to exist, yet there it was—contradicting centuries of theoretical and observational work on how planets form, where they form, and how they evolve with time.

To understand just how strange these planets are, we need a little more background. Since the 1600s, astronomers studied the planetary bodies in our solar system, trying to understand their origins. Improvements in our understanding of physics, and developments in our ability to observe the planets, led to a growing expectation that the solar system shouldn't be unique, that the same rules that governed our beginnings were universal—being in force around the billions of stars across the galaxy.

While lacking the technology to observe a distant planetary system, people have long speculated on their possible existence. In 1584, the Dominican Friar Giordano Bruno published his *On the Infinite Universe and Worlds*, wherein he suggested that the stars in the sky were Suns of their own, and could harbor their own planets. This belief was one of many that brought him before the Roman Inquisition and contributed to his being burned at the stake. While Bruno died, his idea persisted— though it would be nearly four hundred years before his hypothesis of distant worlds would be vindicated.

In the 1700s, Emanuel Swedenborg, Immanuel Kant, and Pierre-Simon Laplace theorized that a star, like the Sun, could form from the collapse of an enormous cloud of gas, or *nebula*. If that initial cloud started with a small amount of rotation, some of the gas would flatten into a disk of circulating material surrounding the star, much like the dress of a spinning dancer flattens into a disk. Dense material would settle to the midplane of the disk, and would eventually coagulate into a set of planets.

Over the subsequent two centuries, this *nebular theory* of planet formation was tested, revised, and tested again using observations of the planets in the solar system. This simple theoretical premise provides a consistent explanation for a lot of the properties that we observe. For example, it explains why all of the planets in the solar system orbit in roughly the same plane—a fact known to humans since prehistoric times. The theory predicted that all of the planets would orbit the Sun in the same direction, as was also known. The nebular theory explains these trends as a consequence of the flat disk from which the planets formed and its rotation around the Sun. Subsequent discoveries in the solar system, like the 1781 observation of Uranus and the later discovery of Neptune in 1846, held true to these predictions.

The nebular theory also explains why Jupiter emerged where it did and how it got so big. High temperatures in the inner solar system prevented volatile compounds that are rich in hydrogen from condensing to make planets. Those same materials could condense out where the giant planets are located, giving more food for the growing Jupiter to feed upon. This also explains why Earth, Venus, Mercury, and Mars formed when and where they did, and why they are made from heavier stuff. The nebular theory had an excellent track record when it came to explaining what we saw in the solar system.

Until recently, virtually all of our knowledge about planet formation was based on observations of the solar system, since that is all that astronomers were able to see. Nevertheless, nothing in the story of our origins was deemed to be particularly unique to the conditions surrounding our Sun. We expected the essential elements of our theories to apply almost everywhere, and there was every reason to expect planetary systems much like ours to form around distant stars, since they would be subject to the same physical principles. The discovery of 51-Peg showed that perhaps these carefully crafted rules may not apply to other stars after all—that either their formation, or subsequent dynamical histories, diverged from the prevailing paradigm.

Many of the early discoveries of exoplanets ran counter to key predictions of the nebular theory. Planets with the mass of Jupiter, and presumably made of the same gaseous material as Jupiter, were supposed to form out in the frigid hinterland of the system, not right next to the star, where it would be subjected to intense radiation. Yet as new exoplanet discoveries accumulated, they continued to defy explanation. Even today, thirty years after they were first seen, we are still trying to understand the origins of hot Jupiters.

Exoplanet discoveries, while concentrated in the last few decades, were enabled by several technological advances that occurred over the last few centuries. One instrument, in particular, that emerged as a capable workhorse was the spectrograph. First employed in the late 1800s, spectrographs take light from a distant source, and break it into its spectrum of constituent colors, like a prism spreading the light from the Sun into a rainbow. When mounted on a telescope and pointed at a star, the spectrum of light from that star similarly spreads into its array of colors.

We can learn quite a lot about stars with a spectrograph because, when light from either the Sun or a distant star passes through the upper layers of the star's atmosphere, it shows a pattern of dark spaces or gaps called *spectral lines*, where specific colors are missing. These spectral lines arise because of the different chemical elements in the stellar atmosphere. The structure of the atoms of an element, or the structure of the molecules of a compound, has a unique set of energy levels for the orbiting electrons. (Imagine each element or compound having a ladder of states, where the positions of the rungs of the ladder are different for each substance—like a fingerprint for that material.) The elements in the stellar atmosphere absorb the specific wavelengths of light that correspond to those energy levels.

This absorption causes the dark spectral lines, and allows scientists to measure the star's chemical composition. For example, in 1868, Jules Janssen and Joseph Lockyer independently used a spectrograph to discover a new element in the atmosphere of the Sun. After shining the Sun's light through their instrument, Lockyer identified the fingerprints of several known elements, along with one set that hadn't been seen before. He named the unknown substance responsible for this new fingerprint "helium" after the Greek word for the Sun—"Helios."

As spectrographs improved, astronomers used them to study the properties of distant stars—learning about their composition and how they compared to the Sun. Spectrographs became so precise that new, more subtle signals could be gleaned from the stellar spectra. If the star was moving relative to the Earth, then the wavelengths of light would be stretched or compressed by that motion. This is exactly the same phenomenon, the Doppler effect, that causes the sound from a passing siren to produce a high pitch with shorter sound waves when it approaches, and a lower pitch with longer sound waves when it recedes. Here, using a spectrograph to measure the stellar spectral lines, and comparing those lines to what we see in a laboratory setting, we can measure how they are stretched or compressed by the Doppler shift. This shows how the stars move along our line of sight, or their *radial velocity*.

Initially, we saw only the average, large-scale motion of the stars in the galaxy, but as these measurements became more and more precise, astronomers started to see slight variations in the speed with which stars moved—periodic shifts superposed on their otherwise constant motion. The cause of these variations was the presence of companion stars that orbited the primary star. As the two stars moved about each other, the Doppler effect would cause the spectral lines from those stars to periodically shift from longer to shorter wavelengths and back.

Astronomers had long known that many stars in the sky were actually multiple stars that orbited each other—they could trace the orbits of the stars in the sky. Some stellar pairs take centuries to circle each other, and the wide separation of these stellar binaries could readily be seen using the telescopes of the late 1700s and early 1800s. Now, with the spectrograph, astronomers could study the orbits with much greater precision. Stellar pairs were soon found whose orbits were only a few days—orbits too close together to see with a telescope alone. This type of system was unknown until the spectrograph uncovered its existence.

As the precision of our spectrographs improved, our ability to measure the Doppler effect caused by the orbits of smaller and smaller objects also improved. It reached a point in the 1950s where the Russian–American astronomer Otto Struve proposed using a spectrograph to detect planets orbiting distant stars. Planets are thousands of times less massive than stars, so the Doppler signal from one would be a thousand times smaller than what was seen for binary star systems. In a 1952 article in *The Observatory* entitled "Proposal for a Project of High-Precision Stellar Radial Velocity Work," Struve noted that the cutting-edge spectrographs would be capable of detecting Jupiter-mass planets if those planets happened to orbit close to their host stars, with orbits of roughly one day [2]. Planets that massive and orbiting that close could cause a Doppler signal large enough to be seen as periodic shifts in the wavelengths of the dark lines of the stellar spectrum.

His claim was largely true on the technology side, but the premise of the proposed observations was shaky on theoretical grounds. Everyone "knew" that Jupiter-like planets wouldn't exist that close to the host star. The nebular theory predicted that they would form at large distances, where it was cool enough for hydrogen-rich compounds like water, methane, and ammonia to condense. Our theories gave no reason to expect a Jupiter-mass planet to orbit its host star in one day. Even if we were able to see its tiny Doppler signal, it shouldn't be there in the first place. Struve didn't propose a mechanism to form such massive planets on these short orbits. He simply argued that since some binary stars were seen to have orbits that small, planets might have them too. He turned out to be right—such planets indeed exist, notwithstanding the predictions of planet-formation theory.

It would take a few decades, but eventually our observations caught up to Struve's imagination, and his approach produced most of the first exoplanet discoveries. His speculative paper has been cited over a hundred times in the scientific literature—over half of them in just the last ten years, and all but six of them since the search for exoplanets began picking up steam in the 1980s. By the late 1980s and early 1990s, there were a few telescopes around the world with spectrographs powerful enough to measure a star's motion to within a few meters per second.

This is an incredible level of precision. Imagine being able to observe a professor pacing back and forth in front of a class ten miles away, and measuring the speed of the motion by looking at changes to the color of the laser pointer in their hand. Or imagine looking at a star that is a hundred times the size of the Earth, located a trillion miles away, and seeing it move at the speed of a person leisurely strolling down the sidewalk. These instruments are pretty sensitive, and coupled with a few years of observations were good enough to start harvesting the low-hanging fruit—systems that cause the most significant Doppler effects on the host star. For exoplanets, that fruit is giant planets on short orbits.

Throughout the 1990s, following closely on the heels of the discovery of 51-Peg, the Swiss were joined by a collection of American astronomers from California, Texas, and Massachusetts (all working independently). Planet discoveries started gracing the pages of scientific journals and newspaper headlines. Each new planet was more weird than the last. Some were several times more massive than Jupiter, and barely classified as planets. Others had highly elongated, *eccentric* orbits that plunge toward their host star, passing within distances only a few times larger than the star itself, before being flung back out to the hinterland. Had these planets been in our solar system, they would have crossed the orbits of all the inner planets—eventually smashing into them or ejecting them from the system altogether. On top of this madness, all these newly discovered planets were orbiting too close to their stars.

The observations were like a parade of counterexamples to the predictions from planet-formation theory (namely that small planets should be near the host star, with large ones more distant, and all on circular orbits). Clearly, the universe of possibilities was larger than anticipated. Despite these discoveries, astronomers were loath to throw away the nebular theory, not because of some nostalgic affection for it, but because it did such a good job matching the solar system, and because it only relied on a few, unremarkable assumptions. The theory's assumptions were so generic, and predictions so straightforward, that they should be seen virtually everywhere—except, apparently, everywhere we looked.

As the number of exoplanets mounted, each piece of information we could extract was certain to provide new insights into their formation and subsequent history, and how their past was different from the solar system. Some of the most valuable information about planets is the sizes and shapes of their orbits, and the sizes and masses of the planets themselves. The Doppler measurements of these systems can determine the orbital size and shape, and also can give a rough estimate for the planet masses. But, in order to really compare exoplanets with the planets in the solar system, we needed to measure their sizes—specifically, their radii—and spectrographs simply can't provide that information. We need a different kind of measurement to do that.

A planet's radius, if it can be found, gives you a lot of information about the planet. You can find the planet's volume, and with the volume you can determine its density. The density measurement gives insights into the types of materials that compose the planet. For example, a large volume for a planet with a given mass implies that it is made of lighter material. On the other hand, a small volume for the same mass requires more dense material. Low-density planets, like Jupiter and Saturn, would be made primarily from gases or light elements. High-density planets, like the Earth or Mercury, would be made from rocks and metals.

Despite a variety of potential detection methods, distant planets are just too small and are hard to see outright. This limits what we can learn about them since, in order to know what planets are made of, we still need a way to determine their sizes. In the year 2000, we got what we needed. Two teams of astronomers observed a planet that happened to pass in front of its host star. That is, the planet *transited* the star. As it passed, the less-than-memorably-named HD 209458b, blocked a portion of the starlight. This planetary transit was a small signal—the star changed its brightness by only one percent as the orbiting Jupiter-sized planet swept across the stellar disk—but it was an unmistakable blip that stood out from the ever-present noise that appears in any observational data.

Planetary transits, while rare, happen regularly in the solar system. In the early 1600s, Johannes Kepler published his work on the orbits of the solar-system planets. He predicted in 1608 that both Venus and Mercury would transit the Sun, and would be visible from the Earth. It was around the time of his prediction that the telescope was invented—precisely the type of instrument that would enable this kind of observation.

Within a few years, Mercury was seen to transit the Sun for the first time, followed shortly thereafter by a transit of Venus. Seeing Venus transit was fortunate timing, since it would not transit again for another hundred years. Mercurial transits happen about once per decade. Venusian transits happen in pairs, about once per century, with the most recent pair being in 2004 and 2012. (The next time will be in 2117—so if you want to see it when it happens next, be sure to eat well and exercise.) Observing these transits of solar-system bodies based on Kepler's predictions was a remarkable feat. While humans had known of both Mercury and Venus for eons, tens of thousands of years passed before anyone saw them silhouetted against the Sun.

Despite having observed planetary transits in the solar system for nearly four centuries, the transit of HD 209458b in the year 2000 marked the first time we had seen this effect outside the solar system. These transits allowed us to measure the radius of that planet because, during the transit, the amount of light it blocked was equal to the relative sizes of the star and the planet. Given the size of the star, which we can estimate from computer models, we can determine the size of the planet that is transiting that star—the crucial piece of information missing from Doppler and other measurements.

This transit measurement was a big deal. It opened a completely new arena for investigation. By combining both the size measurements from the transits, and the mass measurements from the Doppler shift, we can get an idea of what materials compose the planets—whether the planets are rocky or gaseous, or somewhere in between. This new transit information allows direct comparisons between the sizes and masses of solar-system planets and the newly discovered extrasolar-system planets that are orbiting distant stars.

Astronomers had considered transits as a possible exoplanet signature for decades—indeed, Struve's 1952 paper outlining the use of spectrographs to find planets also mentioned planetary transits. However, as with any detection method, there are technical challenges one must overcome. For one thing, as we saw with HD 209458, the signal is quite small. A planet like Jupiter, which is one-tenth the size of the Sun, only blocks about one percent of a star's light (a planet with one-tenth the radius of the star would have an area that is a hundred times smaller). This relationship makes the transit signal much smaller for a smaller planet. The Earth, for example, is ten times smaller than Jupiter and would block only one-hundredth of one percent of the light from the Sun. This fact makes finding Earth-sized planets particularly hard because there are other astrophysical effects that muddle the search. There are spots on the Sun's surface that have a larger effect on the Sun's brightness than a transit of the Earth would have. Nevertheless, despite how small this change in brightness would be, from the 1960s through the 1980s transits were deemed a viable method to find planets given the technological capabilities of the time—except for another small problem with this method: we would need to be exceptionally lucky.

Detecting a transit requires that we somehow catch the planet in the act. The chances of having the orbit of a distant exoplanet coincide with our line of sight is already small. A random view of the solar system would line up with the Earth's orbit only about one percent of the time. For Jupiter, whose orbit is five times larger than the Earth's, a chance alignment would only occur one-fifth as often. The unlikely geometrical alignment is only the first strike against this method. We must also consider that Jupiter-like planets (the planets easiest to see) in Jupiter-like orbits (the orbits where our theories say we should find them) only transit once per decade, and for only a few hours.

A transit of a Jupiter-like exoplanet is an incredibly rare event, and designing a campaign to find even a single example led to some prohibitively low probabilities. To realistically capture a planet like Jupiter in the act of transiting its host star, you need to choose a target star. You hope that it has a planet. You hope that the planet is in one of the 0.2 percent of orbits that will geometrically align with the Earth and therefore will be seen to transit. Then you stare at it, without blinking, for ten years, looking for a one-percent change in the star's brightness that lasts a few hours. Sound fun?

Fast forward to the late 1990s, where the discovery of hot Jupiters significantly changed the detection probabilities. Hot Jupiters are much more likely to transit (about ten percent of the time instead of two-tenths of one percent). They also transit much more frequently, roughly once per week instead of once per decade. Had their existence been known twenty years earlier, astronomy in the 1980s and 1990s may have looked much different. But these kinds of planets weren't known until we unexpectedly stumbled across them, so the solar system was our only example to work from, and the properties of the solar system drove both the expectations and the design of campaigns to find exoplanets.

In the 1970s, the NASA Ames Research Center in California's Bay Area, held a series of seminars to discuss the possibility of finding distant planets. Here, the idea of a large brightness, or *photometric*, survey for transiting exoplanets drove a NASA engineer, William Borucki, to investigate the possibility by examining both the pitfalls and the potential of the approach. Having previously worked on the Apollo missions, he felt this was a good challenge to accept in their aftermath. In 1984, he published his first paper on the subject, and began a series of workshops to identify the best technology to use as the primary detector [3].

There are several devices that can be used to measure the brightness of stars, but these *photometers* tend to be bulky, they often need a fiber optic cable to connect them to the telescope, and they usually have to be cooled to reduce the noise in the data they produce. Observing a lot of stars would result in a Medusa-looking system that would be nearly impossible to make work. However, there was at least one new technology developed throughout the 1970s that appeared promising—the digital camera.

Digital cameras use charged-coupled devices, or CCD chips, which are arrays of tiny semiconductor pixels. Each pixel can store electrons that are dislodged from the surrounding material when photons of light strike the surface. Those electrons are then "read out" and counted in order to determine how much light arrived at each pixel's location during the observation. With large CCDs, you can watch lots of stars continuously, and do so for a really long time. CCDs, if used to search for exoplanet transits, would allow you to make up for the rarity of a transit by providing a way to study more stars at once.

Borucki suggested that a digital camera, attached to a large telescope and lofted into space, could continuously observe tens of thousands of stars, recording their brightness every few minutes. Being in space meant that the confounding effects of the atmosphere would be eliminated, and he could detect the transits of smaller planets. Smaller planets were expected to be on smaller orbits, so he wouldn't need to rely on chance transits of the slow, cumbersome orbits of larger gas giants—he could look for Earth-sized planets that had one-year orbits. Instead of needing to observe for several decades, he would only need to look for a few years. If he observed enough stars at once, the sheer number of targets would overcome the small probability of a chance alignment, and the continuous monitoring—which is more easily done from space—would ensure that none of the transits slipped through the cracks.

Like most new ideas, people thought this was crazy. Despite the success of small projects to demonstrate the capabilities of the technology, there were regular calls to put a stop to Borucki's work. After all, this was five years before the launch of the Hubble Space Telescope, ten years before the first exoplanets were discovered, and fifteen years before one was seen to transit. To many, his work was clearly a waste of resources, with one NASA lawyer even questioning its legality. He was given one last opportunity to make his case. NASA assembled a panel of experts to review his idea—chaired by Jill Tarter, who was known for her work on searches for extra-terrestrial life. William was told to either convince them, or to quit. When he finished, a few of the panel members, notably including Jill Tarter and Gibor Basri (a stellar astrophysicist from Berkeley), joined his team.

Despite the green light to continue working on the idea, he faced another challenge from the fact that there was no good way for NASA to fund the whole enterprise. At the time, NASA basically had small "explorer" missions, and large, strategic or "flagship" missions—there was no middle class for ideas of the size and scope that William had in mind. Without the ability to apply for the right-sized pot of money, his idea couldn't get further off the ground than small research projects to test various technologies. However, this state of affairs changed in the 1990s when NASA unveiled its *Discovery Program* for mid-sized missions. This was a program where the available funding matched what would be needed for Borucki's idea. So the group of fellow travelers that he had accumulated over the preceding half decade started their design of the FRESIP mission, an abbreviation for Frequency of Earth-Sized Inner Planets, which would hopefully fly as one of these new discovery-class missions [4, 5].

As with most scientific proposals that appear on NASA's desk (speaking from experience), this one was rejected. It was too risky.

There was hardly any justifying ground for the FRESIP mission to stand on. At the time, people thought that, while they may exist, planets were probably rare and the best ones to find had longer orbits than the duration of the proposed mission. Besides, the technical challenges to detecting an Earth-sized planet orbiting a Sun-like star in a one-year orbit were substantial. How could you know the camera would function well enough to distinguish such a small signal? How could you know that there would be enough planets out there to find the one in a hundred that would pass across the line of sight? How could you be certain you weren't just looking at star spots or other astrophysical events?

The reviewers of the proposal indicated that it would have been the highest-ranked proposal, but they had serious doubts about the untested CCD technology. Before NASA would take the proposal seriously, William had to show that it was possible to do what he proposed. He kept shopping FRESIP around to scientists and engineers both at NASA and at other institutions, adding new people with new expertise to the team, while NASA continued to demur. Over time, a handful of discoveries from different parts of the world would slowly change William's fortunes.

In the early 1990s, Aleksander Wolszczan and David Frail were using the giant Arecibo observatory in Puerto Rico to study pulses of light emanating from a rapidly rotating corpse of a dead star, a so-called millisecond pulsar. Wolszczan found that the timing between pulses coming from the star was changing—spreading apart, then coming back together, then spreading apart and coming back together again. This should not be. Pulsars, especially millisecond pulsars, emit exceptionally stable signals by astronomical standards. In fact, they are regularly used as astronomical clocks. (A set of fifteen pulsars, each with its spin period indicated, was used as a way to pinpoint the location of the Earth on the famous plaques that were attached to the Pioneer and Voyager spacecrafts. Pulsars were also used to discover distortions of spacetime that span huge distances across the galaxy by watching for small changes in their pulsation frequencies.)

While the pulsation periods of pulsars do drift over time, that evolution is gradual and (except under specific conditions) it always slows down. The observed periodic variation was something not seen before, or predicted. Plus, it wasn't just one variation. The time between pulses was fluctuating on two different timescales. Attempts to construct a theory to explain these variations pointed in one direction—planets. Wolszczan had discovered a system of two planets orbiting this pulsar, the first planetary system seen outside the solar system [6].

These pulsar planets were found in 1992. Then came the discovery of 51-Peg in 1995, the first exoplanet orbiting a Sun-like star. This distinction is an important one, as the planets that orbited the pulsar likely formed after the demise of the progenitor star, while the planet orbiting 51-Peg has a formation history that compares more directly to our own. The pulsar planets notwithstanding, 51-Peg is typically hailed the first exoplanet, as seen by the fact that its discovery was awarded the 2019 Nobel Prize in Physics. Nevertheless, as strange as they were, the pulsar-orbiting planets broke a lot of ground for the field of exoplanets.

The pulsar planets also boosted the motivation to look for other planets. William and his ever expanding team (this time including the great Carl Sagan) revised and submitted the FRESIP proposal again in 1994. The rejection this time was because it was deemed too expensive. Then in 1996, rejected, but at least it was "highly meritorious." In 1998, rejected.

With each rejection came a new list of questions and concerns to be addressed. William and his team needed to show was that it was possible to search for planetary transits by measuring the brightness of tens of thousands of stars simultaneously. With that many targets, spread over a wide field, all sorts of false signals could creep into the data. The mission was supposed to simultaneously look at roughly a hundred thousand stars and find changes in brightness of fewer than a hundred parts per million on each one—which, when they happen, only lasts for a few hours each year. What William was proposing to do would be like looking at Las Vegas from space and detecting a fly buzzing around a streetlight.

To show that this task was possible, he and a group of collaborators made a small, wide-angle telescope at the Lick Observatory in the mountains east of San Jose, California, and pointed it skyward. This *Vulcan* survey didn't find planets, but it did find hundreds of binary stars that eclipsed each other—a signal similar to a planetary transit. Vulcan showed that the team could take and analyze brightness data for the large number of targets expected for the ambitious space telescope. (A copy of the setup was later built in Antarctica, where the long winter night would provide good conditions for continuous observations.)

The last major request from the NASA reviewers was to show that the proposed telescope and instrument design was sensitive enough to see, under realistic observing conditions and with a realistic instrument, the tiny brightness change that would result from the transit of an Earth-sized planet in a year-long orbit around a distant, Sun-like star. NASA needed a successful end-to-end trial that included the drift that would occur while the spacecraft flew, readout noise from the electronics, varying operating temperatures, flight software, and more. In short, William needed to show that every link in the chain, from the photon detection to the analysis software, would work before NASA would be willing to give him the half-billion dollars he was asking for.

A Kepler test facility with all of the trappings was built by Ball Aerospace, the industrial partner who would ultimately build the spacecraft if it were funded.[1] Here, the mission's second-in-command, the Deputy Principal Investigator David Koch (not of wealthy industrialist fame), devised a clever experiment to show that the proposed mission had the necessary sensitivity to accomplish the task at hand. To prove his point, he needed to make an artificial starfield that matched the brightness of some of the dimmest targets that their mission would observe. After that, he needed to somehow change the brightness of these fake stars by only a few parts per million over the course of several hours. Finally, once he had those fake stars and fake planetary transits, he needed to show that the camera could see the tiny effect. (Plates 1 and 2 show William Borucki and David Koch.)

---

1. As a side note, it was only in 1993 that Ball spun off its century-old canning jar business to focus primarily on aerospace.

They made the starfield by drilling holes in a piece of metal and illuminating it from behind. To mimic the tiny change in brightness from a planetary transit, they stretched a thin wire across the holes. However, it wasn't the presence of the wire itself that made the synthetic planetary signal. Instead, they caused the wire to change size—by a few parts per million—by passing an electric current through it. The current made the wire heat up and expand slightly for each degree it rose in temperature. The expanded wire would block more of the light that would otherwise pass through the holes—producing the small change in brightness that was equivalent to a transiting, Earth-sized planet.

With this setup they ran the test to show whether or not the camera, along with the rest of the hardware and software, was up to the challenge. The camera passed. The collaboration resubmitted their proposal and the mission formerly known as FRESIP, but renamed in 1996 to Kepler, was selected by NASA in December of 2001. It would be the tenth mission for NASA's Discovery Program. Over the next several years, William and his team worked to build the Kepler space telescope and prepare it for launch.

This good news for the Kepler team arrived only a few months after I began my graduate-school career. While they were busy putting their mission together, I was bouncing around between different research groups, and working toward graduation. All the while, planets kept surfacing in the observational data. By the early 2000s, there were enough new results to cause people's heads to turn. More and more astronomers were paying attention to these new celestial objects, including my graduate advisor and me.

With the discoveries of transiting exoplanets, beginning with the detection of HD 209458b, my advisor believed there might be more we could learn about them than just their sizes and orbital periods. Given transit measurements with sufficient precision, he thought we could uncover information about additional planets in the system that other instruments might miss. He went on: If the planet were alone, with no sibling planets, then the planetary transits would occur at regular intervals—once each orbital period. But, if there were a second planet in the system, the two planets would interact with each other gravitationally. These interactions would cause small changes to each planet's orbital period. The changes in the orbital period would show up as deviations in the time interval between successive transits. Sometimes one planet would cause the other to transit earlier than normal, and sometimes it would cause the transit to be late.

He wondered how large these "transit timing variations" might be, and how they might change depending upon the properties and orbits of the other planets in the system. Perhaps we could measure the masses of the planets without the need for expensive, cutting-edge spectrographs (the largest spectrographs can cost millions of dollars to build, and nearly ten thousand dollars per hour to operate, after all). It might be possible for us to detect and characterize unseen planets—ones that didn't transit, or were too small to see or to be deduced with spectrographs. Maybe we could directly measure the size of the star so that we didn't need to rely on computer models when estimating their size and the sizes of the planets that orbit them—as was the case up to that point. Perhaps we could constrain some details about the shape of the planetary orbits. All of this information might come out of an analysis of the variations in the transit timings. My "practice problem," while we waited for me to complete the general relativity course I needed for the black-hole-imaging research, was to explore these possibilities.

We got started working through the pages and pages (and pages) of algebra. Upon finishing a calculation, we did it again, and again. A constant challenge with this line of work is that a single misplaced minus sign, or one mistake in a summation, and the whole calculation is wrong. After a few months we convinced ourselves that we had done the math correctly. We wrote some computer models that matched our calculations, adding to our confidence. As we worked, something became increasingly apparent. The mutual gravitational interactions of the planets in the system would cause changes to their orbital periods that were surprisingly large. Under the right circumstances, a small planet like the Earth could change the orbital period of a large planet like 51-Peg or some other hot Jupiter (which are over three hundred times heavier than the Earth) by minutes, or even hours. That was a big signal.

With the existing, ground-based technologies, astronomers could measure the moment that a planet transited its host star to within about one minute, so a deviation of an hour would be easy to spot. Through our work, we had indeed uncovered a way to find hidden planets, and our method was exceptionally sensitive. These transit timing variations, soon shortened to the acronym TTVs (the creation of which may be my greatest contribution to exoplanet science), were possibly the best method available to find Earth-mass planets orbiting Sun-like stars. No other options available at the time really compared to TTVs—neither Doppler measurements nor existing transit measurements could find such small planets. We submitted our initial paper on this method for peer review in late 2004, in conjunction with a similar study from Harvard, feeling good about the likely value of our work [7, 8, 9]. To use this technique, however, we would need the right data to analyze.

More than a dozen transiting exoplanets had been seen by this time, but the kind of data that existed in the scientific literature wasn't good enough for us to make use of our methods. Most systems had only one or a few transit measurements, and we needed a series of lots of measurements. Just as we wrapped up our initial work, a group of astronomers published nearly a dozen observations of a transiting planet—a planet called TrES-1b. These were several, good-quality transit times for a single system, spread over a large time interval. They were precisely the data we needed to try our hand at finding unseen planets with TTVs.

With our paper undergoing peer review, and likely to be accepted, we had a decision to make. Do we keep pursuing research involving exoplanets and TTVs? Or do we turn our attention back to imaging supermassive black holes?

When a theorist in physics or astronomy announces a potential signal, such as ours, they often need to wait for years or decades before adequate data are available to search for that signal. The data on TrES-1 appeared at just the moment when we were ready to use them. My advisor said to me, "It isn't very often that the data you need fall in your lap." With these data and our methods, we had a chance to be the first people to detect an Earth-mass planet orbiting a distant star. It was an opportunity not to be missed. With that, we dropped our plans to image black holes, and shifted our efforts entirely to exoplanets. My so-called practice problem became the subject of my PhD dissertation, and over the next two years I would build the methods and software that I needed to analyze transit data. I could then apply them to transiting exoplanet systems, looking for hidden planets in those systems as new observational data started to appear. (For the record, in April 2019 a group of scientists did take that black-hole image using the Earth-sized Event Horizon Telescope, but by then I was far down the exoplanet rabbit hole.)

Now the stage is set. By 2006, planet-formation theory was in shambles as the discoveries of the previous decade kept defying its predictions. In order to understand what was happening, the scientific community needed more data to shed light on the issue, and it would require more than just a few new discoveries here or there— they needed a large, comprehensive survey. After two decades of development, Kepler had been selected for construction and flight. The growing Kepler Science Team now included many of the scientists whose exoplanet discoveries had made the headlines of the previous decade. The engineering team was actively building and calibrating the instrument. The software engineers were coding the analysis tools needed to find planets in the flood of data expected from the spacecraft. And the astronomers were working feverishly through the enormous pile of preparatory work required to make the mission succeed— cataloging, characterizing, and selecting the stars that the telescope would observe.

The mission was also looking for new blood to join the science team. They wanted people whose research would provide novel insights into the discoveries from the Kepler data, or perform some analysis that wasn't envisioned when Kepler was first designed nearly two decades in the past. This request included a specific reference to the "detection of non-transiting planets by timing of the variations of the transit epochs." I had gone to the University of Washington, where my career preconceptions were dead on arrival, but I had just defended my practice-problem-turned-dissertation, "Detecting New Planets in Transiting Systems," by analyzing the timing of the variations of the transit epochs caused by the mutual gravitational interactions of planets in those systems [10]. My research had been about analyzing precisely the kind of data that Kepler would produce, and at that fleeting moment, I was the only person on Earth who had done so—twice. In short, I won the scientific lottery.

NASA had just asked for someone like me to join their science team. I was a newly minted PhD, working at a national laboratory that let me submit a proposal to the competition. And, when the powers that be at NASA accepted my proposal, I was given the opportunity of a lifetime to work with the best colleagues, in the most exciting scientific field, on the hottest mission in the world. Kepler would revolutionize astronomy, I had a seat on the ride, and was about to have the trip of a lifetime.

2

Designing the Kepler Mission

Shortly after the invention of the telescope, through a series of observations he made in 1610, the Italian astronomer Galileo began to establish the physical nature of planets as spherical objects orbiting the Sun. For the thousands of years prior to that time, planets were generally conceived as unusual star-like objects with only the Earth, Sun, and Moon being recognized as heavenly spheres. (Yes, despite modern advocates for the contrary, we have known the Earth is a sphere for at least two millennia.) What sets the planets apart from the stars is the fact that they don't stay in the same place in the sky. The pattern of stars we see in the sky remains fixed over the course of a human lifetime—their tiny relative motions can only be seen with the most powerful telescopes. The planets, however, drift relative to the constellations of stars over the course of only a few nights. From one year to the next, their persistent motion can carry them across the entire sky, passing through constellation after constellation. This continual movement of the planets is what gave them their name—the word "planet" derives from the ancient Greek "plánētes astéris," meaning "wandering stars."

One early mystery about planets is that, occasionally, some of them pause their motion relative to the background stars, back up for a few weeks, then proceed along their original path. For centuries, predicting these *retrograde* loops made by the planets was a central focus of astronomers. They devised various theories for how, why, and when these loops occurred. Aristotle, and later Ptolemy, thought that the Earth was at the center of the universe with the planets, attached to some heavenly sphere, circling around the stationary Earth. These heavenly spheres explained the general motion of the planets across the sky, while the retrograde loops were caused by a second sphere, called an *epicycle* attached to the first. As the two spheres rotated, one centered on the Earth, and the other centered on the edge of the first, the path of the planet would trace periodic loops in the sky, like the patterns traced by a pencil in a spirograph.

This Earth-centered model for the heavens prevailed in the ancient Greek, Roman, and later European cultures. However, several Greeks, notably Aristarchus of Samos, supported the idea that the Sun was at the center of the solar system, with the planets orbiting around it. This Sun-centered model lost favor among scholars for the better part of two millennia until, in the 1500s, Nicolaus Copernicus reintroduced it and used it to explain planetary retrograde motion. Here, the planets orbit the Sun at different speeds, with planets closer to the Sun moving faster. When the Earth passes a slower-moving planet, the planet briefly appears to move backward, making a retrograde loop, even though its motion remains constant relative to the Sun. This is much like the situation where a car with a perpetual driver-side blinker, traveling at 55 miles per hour in the fast lane of the freeway, appears to move backward as you speed by them on their right. Their speed remains constant regardless of the number of cars that back up behind them, but as you pass them you would need to turn your head if you were unable to resist the temptation to glower condescendingly at them as they recede from your presence. In the solar system, as we pass by the outer planets due to our faster motion, they also appear to move backward in the sky for several weeks.

This Copernican model still had some flaws, but by the first decade of the 1600s, the observational evidence strongly favored the Sun-centered model. The bulk of the evidence to support the Sun-centered solar system came from a cutting-edge observatory that had been built on a small island between Denmark and Sweden. Since telescopes didn't exist during the last quarter of the 1500s, sightings of planets and stars were made by eye. New techniques to correct for the effects of the atmosphere, and large instruments, some being several meters in size, were used to increase the precision of the observations. Over many years, a catalog of a thousand stars and the changing positions of the planets were tabulated by the Danish astronomer Tycho Brahe—the ultimate naked-eye astronomer. The number of new observations taken by Brahe dwarfed what had been documented since antiquity by European and Middle Eastern astronomers.

After Brahe's death in 1601, his data were compiled and analyzed by his assistant, Johannes Kepler. Kepler (for whom NASA's Kepler space telescope was named) used Brahe's observations to formulate his three laws of planetary motion. The first two of his three laws were published in 1609, with the third coming a decade later in 1619. Kepler's first law states that planets orbit the Sun in an ellipse, with the Sun located at one focus of that ellipse.[1] This law was a significant departure from the historical assumptions about planetary motion. For centuries, all heavenly motion was assumed to follow circular paths—or paths that could be constructed from motion along a set of circles. Kepler's calculations showed that, despite its aesthetic appeal, the circular motion was unable to describe Brahe's findings.

Kepler's second law states that a planet's speed changes throughout its orbit—another groundbreaking claim. The speed of the planet would vary such that, when the planet was closer to the Sun it would move more rapidly, and when it was farther from the Sun it would move more slowly. This changing speed occurred in a predictable way. Specifically, if you were to draw a line connecting the planet to the Sun and you were to specify an interval of time, the area that line would sweep through in that time interval would be the same regardless of where you started in a planet's orbit, and regardless of the length of time you considered. Figure 2.1 shows a diagram of Kepler's first and second laws. With Kepler's work, it seemed certain that planets orbited the Sun rather than the Earth. Now, not only were their orbits different from what was previously expected, and their speed through the

---

1. Each ellipse has two foci. The ellipse itself is the collection of points where the total distance between a point on the ellipse and each focus is a constant. A circle is just an ellipse where the two foci lie on top of each other.

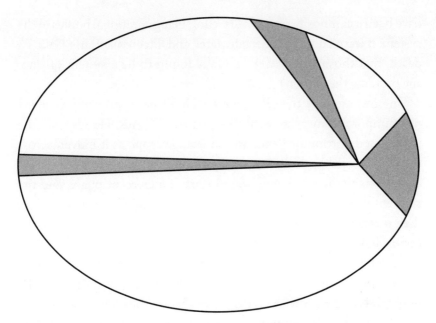

FIGURE 2.1. Kepler's first law states that planets orbit the Sun in an ellipse, with the Sun at one focus (the point where the various lines originate in this figure). Kepler's second law states that a line connecting the Sun and the planet will sweep out an equal area in an equal amount of time. All of the shaded wedges correspond to the same time interval in a planetary orbit. Each shaded region is the same area as the others.

heavens variable, but Kepler's work also implied that the planets may be similar in size and shape to the Earth (round, and with a diameter not too much larger or smaller)—a fact definitively established by Galileo.

The same year that Kepler published his first two laws of planetary motion, Galileo built his first telescope—an improvement on a design that originated from the Netherlands the year prior. The instant Galileo lifted his telescope toward the heavens, he rendered obsolete Brahe's world's-best, cutting-edge observatory. He used this telescope to look at the stars, the planets, the Moon, and even the Sun, making a number of striking discoveries: craters and mountains on the Moon, spots on the Sun, moons orbiting Jupiter, and strange shapes protruding from the sides of Saturn. (Galileo thought that Saturn may have been a composite object—its "ears" being additional bodies in the system. It wasn't until four decades later, and fifteen years after Galileo's death, that those shapes were actually found to be a system of rings surrounding the planet.)

The invention of the telescope enabled what is arguably Galileo's most important measurement, the phases of Venus. He saw that the illuminated portion of Venus would change shape as it moved across the sky—much like the illuminated portion of the Moon changes shape as it orbits the Earth. Venus would start in a crescent phase, and the crescent would grow until it passed through a full phase. The process then reversed until it became a crescent again, this time curving in the opposite direction.

This set of observations showed several new truths about planets. Planets were round. Before Galileo, the planets simply appeared to be bright dots in the sky with no definitive shape to be seen. After Galileo, they were known to be roughly spherical objects, akin to the Earth and Moon. Galileo also saw that Venus was relatively large when it was in a crescent phase and progressively got smaller as it approached the full phase. This showed that its distance from the Earth was changing dramatically. However, his most interesting observation was that Venus had a full phase in the first place. Had Venus been orbiting the Earth according to the prevailing theory of the time, then it would always appear as some form of a crescent phase, with most of the sunlight shining on the side hidden from our sight. The only way that Venus could be fully illuminated by the Sun was if it was on the opposite side of the Sun from our point of view. In other words, Copernicus and Kepler, and Aristarchus before them, were right. With the observation of Venus approaching a full phase, Galileo showed conclusively that the planets orbited the Sun.

There were also a handful of other astronomers, contemporaries of Galileo, who used similar technology to make similar observations and who reached similar conclusions. For the next few centuries, improvements in the telescope produced more discoveries of solar-system objects. Several moons of Jupiter and Saturn were seen by the end of the 1600s—including Jupiter's four large *Galilean satellites* seen by . . . Galileo. We learned that Saturn's strange protrusions were actually a system of multiple rings. A century later, in 1781, William Herschel discovered Uranus—the Sun's seventh planet. Two decades after that, Giuseppe Piazzi discovered the eighth planet, Ceres, located about halfway between Mars and Jupiter. Then three more planets were identified in quick succession: Pallas in 1802, Juno in 1804, Vesta in 1807, and the twelfth planet, Astraea, appeared in 1845.

These last several planets are less famous than their more popular counterparts like Jupiter and Mars. Just because we don't talk about them much doesn't mean that they don't exist—they are real, and they are still there orbiting the Sun. However, there were some key differences between these new planet discoveries and the previously known planets, now including Uranus. For starters (aside from Uranus), all of these new planets orbit at roughly the same distance from the Sun, between Mars and Jupiter. They were also too small for astronomers to measure their size. The other planets in the solar system are large enough that early telescopes could show that they were round. These new planets only appeared as points of light in the sky; they looked more like stars—which also appeared as simple points of light when observed with the telescopes of the time. William Herschel proposed that they be called *asteroids*, or "star-like." Over the next half century, and as new discoveries of similar objects piled up, the distinction between asteroids and planets became more clear. Their sizes were much different, where they lived in the solar system was different, the orientation and shapes of their orbits were different. The asteroids were slowly demoted from bona fide planets to small bits of rocky space debris.

Before that naming convention had settled in, however, a thirteenth planet appeared, and its discovery was different from all previous ones. For several years, astronomers had been tracking the orbit of Uranus and found that it disagreed with the predictions of both Kepler's laws of planetary motion and Newton's law of gravity. To explain these deviations from the models, a French astronomer and mathematician, Urbain Jean Joseph Le Verrier, suggested that the discrepancy might be from the influence of a large planet living beyond the orbit of Uranus. He sent his prediction to Johann Galle and Heinrich d'Arrest at the Berlin Observatory. The night that Le Verrier's prediction arrived (September 3, 1846), they turned their telescope toward the spot Le Verrier indicated and discovered Neptune. An English astronomer, John Couch Adams, was independently considering the same explanation and made similar observations a few months later.

Not only did the observation of Neptune complete the list of known planets in the solar system, it showcased the power of applying theoretical predictions to making new discoveries. Here, it was using perturbations to the orbit of one planet to infer the presence of another. It is essentially the same physical process that formed the basis of my dissertation work on TTVs (transit timing variations), and it turned out to be central to many results from the Kepler mission. Perturbations caused by the gravitational influence of unseen planets are also the basis for the current (as of 2023) hypothesis of a new planet existing beyond the ice giants in the outer solar system—Planet 9 [11]. (For Planet 9, the evidence is the correlation of the shapes of the orbits of several ice balls out in the very distant solar system—well beyond the orbit of Pluto. That correlation may be a consequence of the gravitational influence of an as-yet-undiscovered planet.)

Of course, not all mathematical predictions of a planet turn out to be correct. Mercury's orbit also showed an unexpected deviation from the models. The orientation of its orbit slowly drifts around the Sun. A portion of this drift can be accounted for by the influence of the other planets, and the slight out-of-roundness of the Sun. But there remained some unexplained motion. Le Verrier used his Neptune-finding method again to predict the presence of an additional planet—this time orbiting interior to Mercury. It turned out, however, that the hypothetical planet Vulcan was not there. Clearly, not all theoretical predictions pan out. For the precession of Mercury's orbit, the cause turned out to be a consequence of warped spacetime, as explained by a new model for the theory of gravity: Einstein's general relativity. As for the hypothetical Planet 9 in the outer solar system, we shall see. If it is there, it will be an interesting discovery. If not, we will learn something about the solar system along the way. It's not like we can choose whether or not it is there, only whether or not we look.

Continuing with solar-system discoveries, Clyde Tombaugh announced the detection of Pluto the non-planet in 1930. It had appeared in images taken by various telescopes as early as 1909, but it had not been recognized for what it was. Such oversights are not uncommon. Astronomical objects can often appear in an image, but are not recognized as interesting until years later. Indeed, a similar thing may have happened with Neptune, as there is strong evidence that Galileo actually saw Neptune with his observations in 1612—more than two hundred years before its official discovery date. Other astronomers followed suit in the subsequent centuries by overlooking the presence of Neptune in their observations.

Pluto was counted among the solar system's planets for the better part of a century, but it was always seen as an oddball. It is small, smaller than some moons. Its orbit is tilted relative to the other planets, elongated more than the other planets, and it crosses the orbit of Neptune for twenty years each orbit. Most recently, Pluto was closer to the Sun than Neptune from 1979 to 1999.

As nocturnal surveys of the sky improved through the 1990s, we found several other objects that were similar to Pluto in size, with orbits that were shaped more like those of the planets. These objects were also quite numerous. Astronomers were stuck in the same position they had been the century before with the asteroids. These distant, icy worlds were not really planets, but were some new type of object—dirty snowballs in a belt of debris beyond the orbit of Neptune. Pluto was just one of the largest of the bunch. Unlike the previous century and the slowly changing classification of the asteroids, however, the transition of Pluto from planet to non-planet came rather abruptly. In 2005, a group of astronomers discovered an object that was soon to be called Eris. It too had a highly inclined, highly eccentric orbit. It was also small. However, small as it was, it was more massive than Pluto. Being more massive than Pluto, if we were to be consistent, then Eris should be the tenth planet [12].

By this time these discoveries of similar objects were becoming commonplace, and the possibility of an ever-extending list of solar-system planets was staring us in the face. The international astronomy community got together to bring order to the chaos, to construct a new definition for a planet that could be consistently applied to existing, and new, discoveries. In the end, the definition was essentially what most people expected it would be. That is, a planet is whatever Pluto isn't. Sure, it was couched in more technical language like "hydrostatic equilibrium" and the like, but that was just wordsmithing to get the desired result. Pluto didn't belong on the list, and so the rules were rewritten accordingly. Many people were saddened by this turn of events, feeling sorry for Pluto and its loss of stature. I was not among them. I didn't really care. And to be honest, Pluto probably didn't care either.

The new definition of a planet works for the moment, but the discussion is never closed since new discoveries will require regular refinements in the distinctions between terms. A similar debate still rages on the other side of the planetary scale. How big can an object become before it is no longer considered a planet? There is a class of objects, called *brown dwarfs*, that live in the wasteland between the sizes and masses of stars and the sizes and masses of planets. To some extent, these brown dwarfs are like "the little engine that couldn't." They are not stars since they aren't able to ignite the nuclear fusion that converts hydrogen into helium in their cores. Some can fuse deuterium (an isotope of hydrogen that contains an extra neutron), but deuterium fusion is a lot easier to do than hydrogen fusion, and it doesn't require the high temperatures that one finds in the cores of stars.

Despite not being stars, they appear to form in a manner more similar to stars, and unlike the way planets form. So brown dwarfs aren't stars, but they aren't planets either. Regardless, the dividing line between a planet and a brown dwarf is murky, having a few possible definitions floating around. Some of those definitions, however, are not easy to distinguish just by looking at one from the Earth. For the moment, the ambiguity is just tacitly acknowledged by astronomers, but left unresolved. While this uncertainty doesn't affect the solar system, it does have some implications for exoplanets, as we shall see.

From the early 1600s through the 1980s, the technology used in astronomical observations continued to improve. Each advance brought new opportunities for discovery. In some cases the improvements came in the design of more powerful telescopes. In Galileo's day, a diameter of a few centimeters was all one could use to collect light from distant sources. Today, we are building telescopes that are thirty meters across—a thousand times larger than what Galileo used, with a million times the collecting area. In other cases, astronomical advancement came from more sensitive instruments, or completely new instruments. For example, Galileo's observations were hand-drawn sketches of what he saw. The first photograph (daguerreotype) of the Moon was taken two hundred years later, in 1840. The digital camera made with CCDs appeared in 1969, with the first astronomy-related image taken in 1976. Each of these advancements brought radical changes to the field of astronomy. The development of digital cameras, and the advantages they bring to image sensitivity and data analysis, quickly drove them to displace photographic film as the image medium of choice among astronomers.

For another example, the dispersion of light into its constituent colors had been seen since before modern humans walked the Earth (rainbows are an example of dispersion). Spectroscopy, the study of this effect, took off in the 1600s with the work of several scientists, notably including Isaac Newton. As we've seen, new instruments and techniques brought significant advances in the 1800s, and improvements are still made today. Parallel advances in physics brought improved understanding of the electromagnetic spectrum of light, especially into and beyond the infrared and ultraviolet regions. Heinrich Hertz discovered radio waves in the 1880s, followed in the 1890s with the discovery of X-rays by Wilhelm Röntgen. All of these advances in the science of light, or *optics*, affected what we could observe and how, and led to better understanding of the heavens—from the planets in the solar system to distant stars and galaxies.

Of particular importance for our story was the completion, in 1963, of the Arecibo radio telescope. This 1000-foot dish (305 meters) was initially built to study the ionosphere, a layer of the atmosphere high above the Earth's surface, but Arecibo served a dual purpose as a radio observatory. As we saw in the last chapter, this telescope was used to discover the millisecond pulsar in the constellation of Virgo that hosted two exoplanets (a small, third planet was later discovered in the data). These were the first planets seen outside the solar system. Even now, the third planet in this system is the least massive planet seen outside the solar system (weighing in at only two percent of the Earth's mass). By today's definitions about being massive enough for their gravity to pull them into a round shape, and for them to clear out the neighborhood of their orbit by ejecting nearby debris, their discovery was the first time in nearly 150 years that a new planet was added to the planet list. It also kicked off a decade of groundbreaking planet discoveries—but mostly around main-sequence stars, like the Sun.

As with many scientific results, leading up to the detection of the pulsar planets there were hints of other planetary signals coming from other pulsars. A *substellar* companion was seen to transit a millisecond pulsar in 1988. This companion has an estimated mass roughly twenty times that of Jupiter, a mass that lands it squarely in the ambiguous region between a planet and a brown dwarf. That ambiguity in the definition disqualified it as a definitive exoplanet detection—showing how much these words and their definitions can matter.

While the first set of pulsar planets were being discovered, a number of astronomers around the world were studying the orbits of binary stars using spectrographs. It turns out that multistar systems are quite common. (A famous quote attributed to the astronomer Cecilia Payne-Gaposchkin is that "three out of every two stars are in a binary.") In essence, when you look up in the sky, there is a fifty-fifty chance that any given star is actually two or more stars orbiting each other—you just can't tell them apart with the naked eye. To further drive this point home, the brightest star in the sky (Sirius) is a binary system, the nearest star to us (Alpha Centauri) is a triple, and the most famous star (Polaris, or the North Star) is also a triple.

For decades, astronomers had been measuring the orbits of binary stars, using spectroscopy to measure the periodic shortening and lengthening of the wavelengths of the star's light that are caused by the Doppler shift from their relative motion. That information can be used to determine the stellar orbits and masses. As spectrographs improved over the years, we were increasingly capable of detecting smaller and smaller bodies. By the 1980s, searches for brown dwarfs—the little stars that couldn't—were well established. And the technical capabilities were in place to start finding planets, as Otto Struve had recommended over three decades earlier.

In 1988, after a six-year campaign, promising planet signals turned up around seven stars. Most of those signals were false alarms, but one (Gamma Cephei) was confirmed to be a planet in 2003—after nearly twenty years of observations finally put competing hypotheses to rest. Less than a year after the announcement of the initial seven planet-looking signals, a brown dwarf search turned up another signal orbiting the star HD 114762. That too was later confirmed as a . . . well, with a mass at least eleven times that of Jupiter it was a planet at best, but more likely a brown dwarf. Originally listed as a planet, new observations in 2019 show that it is neither. It's a small star [13].

While stars are clearly not planets, the distinction between a brown dwarf and a planet may not be important in the long run. Regardless, these initial hints of planets are part of the handful of discoveries that predate the first definitive detection in 1995 of the exoplanet 51 Pegasi. As is typical in science, there was steady progress on multiple fronts that led to interesting hints of something new to be found. After grappling with different issues, there is a watershed study, or a flurry of data from a new instrument, that produces what is widely recognized as the groundbreaking result. For exoplanets, this breakthrough moment came from long-term Doppler monitoring of the 51 Pegasi system, discussed in chapter 1. Astronomers saw a signal in their data that unmistakably came from a planet orbiting another star, and that star was like the Sun.

Aside from the improvements in spectrographs that led to these breakthroughs in the field of exoplanets, there were similar advances for other types of instruments. The significant increase in computing power throughout the 1980s and 1990s opened new methods for storing, accessing, and analyzing data. When coupled with digital cameras, the task of finding short, transient events became easier, because the tedious work of studying brightness variations of hundreds or thousands of objects could be assigned to a computer, rather than a graduate student. With lots of digital images, and lots of digital computers to process them, astronomers could quickly analyze them to look for planets.

One example of how digital images and computer processing can find planets comes from a planet's ability to bend light as predicted by Albert Einstein. Suppose you monitor the brightness of a distant star—a very, very distant star. If some object, like another star, passes in front of that original target star, the gravitational influence of the intervening object will deflect more of the light from the target star into your telescope, like a lens focusing more light onto your camera. This brief alignment of the two stars causes the distant, target star to brighten temporarily, while the intervening *lensing* object passes across the line of sight. Typically, the entire *lensing event*, when the distant star brightens and then dims again, takes a few days.

These events are pretty rare because of the unlikely alignment needed for them to occur. Despite the fact that there are hundreds of billions of stars in the galaxy, most stars are incredibly far apart. If you stacked clones of the Sun end to end, it would take twenty-five million of them to reach our nearest stellar neighbor. So, to see a pair of stars lining up with our telescopes means that you either have to be incredibly lucky, or that you have to be looking at a lot of distant target stars to compensate for the low probability of any one of them being lensed by an interloper. Nevertheless, there are a few surveys that look for this effect, and they regularly see the gravitational lensing of one star by another.

For our purposes, these signals get even more interesting if that intervening star happens to have an orbiting planet. In this case you can observe two, and sometimes more, lensing events—one for the star and a second for the planet. While the lensing event for the star may take several days, the smaller *microlensing* event for the planet will only be a couple of hours. The challenge in seeing the planetary lensing signal is that its contribution to the overall event is so short. You have to first identify that there is an initial, star-induced lensing event. Then you quickly tell collaborators around the world to start monitoring the brightness of that target. You have to watch almost constantly, so as not to miss the brief signal from a potential planet, because any hour-long gap in the data is large enough to hide what you are looking for. The only way to accomplish this effectively is with a network of telescopes around the world so that, as the Sun rises in one location or clouds build up in the sky, and nearby telescopes are forced offline, another telescope in a different location can pick up where you left off. So far, there are well over a hundred planets that have been seen with this technique, all enabled by our ability to coordinate observations around the world, take high-quality brightness measurements, and quickly process the digital data.

This planet-detection technique has some downsides. Because the lensing stars are so distant from us, typically being halfway across the galaxy, we often can't identify them, especially since the majority of stars in the galaxy are smaller and dimmer than the Sun. In most cases, when we find a planet through microlensing, we are detecting a planet that we cannot see, orbiting a dim star that we also cannot see.[2] All of the planet discoveries from this method come from the influence the planet and star have on the trajectory of the light from the original, distant, but brighter target star (a strange thing to contemplate).

Two other planet-detection techniques are worth mentioning here before we get back to transits and Kepler. The *astrometry* method tracks the measured position of the stars in the sky. As a planet and its host star orbit each other around their mutual center of mass, the position of the star will shift relative to the rest of the stars in the sky, tracing out the path of the stellar orbit. We can measure the size and shape of that orbit with high-precision measurements of its position as it changes over time. Astrometric measurements have not produced many planets—only two planets are listed in the exoplanet database as of January 2023. But there is currently a mission from the European Space Agency (ESA) called GAIA. GAIA should find

---

2. I'm paraphrasing here a quote that is attributed to Debra Fischer—one of the early pioneers in exoplanet science.

many exoplanets, probably thousands, using this technique. If its results haven't been announced by the time this book is printed, they are expected soon.

The second method is imaging a planet directly. This might also seem a straightforward detection method, just taking a picture and seeing a planet, but the simplicity of the idea quickly faces major technical hurdles. Directly imaging a planet requires seeing light that is reflected from the planet. But the host star is much, much brighter than the planet. For example, the Earth reflects only one-billionth of the light that the Sun emits. So, if you were viewing the Sun from a distance, to see the Earth would require being able to resolve a second source of light that is a billion times dimmer. This is really hard to do, and that is just one of the two main problems that you face when trying to directly image a planet.

The second challenge for direct imaging is that planetary orbits, even at the large distances of many planets in the solar system, appear too close to their stars. If the solar system were viewed from the distance of even the nearest stars, the Earth would look like it was in essentially the same place as the Sun, making it nearly impossible to distinguish between their locations. To see, or *resolve*, both the Earth and the Sun from our nearest stellar neighbor would be like seeing the field of blue stars on the flag waving atop the US Capitol Building, from Los Angeles. This requires extremely high-resolution imaging, by current standards.

Earth's atmosphere presents a big roadblock in confronting this second technical hurdle—resolving two nearby sources. When light from the distant system passes through the atmosphere, it gets deflected by turbulence in the air by a process called scintillation. Scintillation is what causes stars to twinkle. It also causes images to be blurry, increasing the difficulty of separating the source of light from the star, and the source of light from an orbiting planet. To correct for that blur, astronomers use *adaptive optics* systems. Here, there is a small mirror surface in the telescope that can be deformed using a bunch of tiny mechanical actuators. If you measure the effects of the atmosphere, and if you act fast enough (within a few milliseconds), you can change the shape of the deformable mirror surface to counteract the atmospheric distortions.

We measure the behavior of the atmosphere by shining a laser beam into the sky, nearly aligning it with the position of the target star. That laser light reflects off a thin layer of sodium that is high in the atmosphere. When the reflected beam returns to the Earth, it will have experienced the same distortions that affect the light from the distant stars. A computer then measures the distortion of the laser light and sends a signal to the actuators on the back of the deformable mirror. Those actuators change the mirror surface to remove the measured effects of the atmosphere. Only now is the image clear enough to resolve the light from any small planets orbiting around the target star.

Once the atmospheric effects are removed, we can address the first technical hurdle of seeing the tiny planet signal in the presence of the much brighter star. The instrument we use to do this is called a coronagraph. It works essentially the same way that your hand works on a bright summer day in Las Vegas. You hold it out in front of your eyes to block the direct light from the Sun, making it possible to see the light coming from everything else—like the stoplight that you are waiting to see turn green. A coronagraph for astronomical measurements is positioned with a bit more precision than what you get with your hand. It also has a special shape to account for the effects of light scattering off from different parts of your telescope, but the idea is the same. To get it to work well, you need the adaptive optics system just described, and a coronagraph mask that is large enough to block the light from the star, but small enough to allow the light from the orbiting planets to pass by its edge.

Neither of these challenges, the atmospheric scintillation and the glare from the target star, are easy to overcome, but there are many scientists working to improve our capabilities. After a small handful of detections in the early 2000s, a lot of technological development has advanced this technique considerably. Now there are several dozen planets found by directly imaging them. Many of these planets are very young—still cooling from their formation and emitting a glow of their own. Their warmth increases their brightness relative to their host star, making the detection easier. But, in the coming years (or decades), we should be able to see cooler planets using light reflected from their host star.

This brings us to back to planetary transits, the method used by the Kepler mission. Today, transits have produced the largest number of planet discoveries, by a long shot. Finding planets by transit measurements uses a lot of the same technologies described for the other methods—digital cameras, computing power, and continual (or near-continual) coverage.

We've seen that for a transit survey to work, there are three primary obstacles to overcome. The first is that transit signals are small because the amount of light that a planet blocks is proportional to the ratio of the cross-sectional area of the planet and the star. For Jupiter and the Sun that is about one percent, while for the Earth it is one-hundredth of one percent. The second is that transits are rare. You need to survey enough stars to compensate for the unlikely alignment of the planetary orbit with your line of sight. For the Earth, when viewed from a distant star, the misalignment can only be about a half degree before the Earth would no longer be seen to transit—there is only a one-percent chance of the Earth's orbit crossing the line of sight from a random vantage point. Finally, transits are brief. They only occur once per orbit, and they only last a couple of hours. To catch one in the act you need to make new observations every few minutes, a half hour at most, otherwise you could miss the transit altogether. A planet at the Earth's location orbits roughly every ten thousand hours, with a transit duration of about ten hours. So the Earth would only be in transit across the Sun one-tenth of one percent of the time.

To increase the odds, it can help if you already know that a planet is there to begin with. The first exoplanet seen to transit was HD 209458b, which was one of the eleven known exoplanets at the time its transits were detected. Its initial discovery used the same Doppler shift measurements that found the other ten planets. Two teams of astronomers followed up the discovery by observing the brightness of the star in hopes of seeing the planet transit. After a series of observations over several nights, both teams saw what they were looking for—a dip in brightness that was consistent with a planetary transit signal [14, 15].

But we don't want to rely on the benevolence of other exoplanet detections to provide targets for a transit search. So to compensate for the long odds of seeing a planetary transit, surveys monitor large patches of the sky, containing tens of thousands of stars. This is where the digital camera and data processing come in. You set up your telescope and take regular pictures of a part of the sky that contains a large number of target stars. You then analyze all of the stars simultaneously to see whether one of the stars dims by one percent (or less) over the course of a few hours.

In the years leading up to and following the discovery of the first transiting planet, a number of ground-based surveys monitored the sky each night looking for as-yet-undetected planets. Since these surveys concentrated on the brightest stars, whose brightness is easier to measure, the telescopes they used were surprisingly small. It was often sufficient to simply use a camera with a ten-centimeter lens (four inches), or a small array of said cameras. Moreover, a lot of equipment for ground-based transit surveys can be purchased off the shelf. Compare that to the several-meter diameters of the telescopes used to measure planet-induced Doppler shifts of the target stars, and their multimillion-dollar custom spectrographs.

Here are some examples of the surveys that typified the early days of transiting planet discoveries. The Trans-Atlantic Exoplanet Survey (TrES) used a set of three ten-centimeter telescopes. WASP, the Wide Angle Search for Planets, is an array of eight cameras of essentially the same ten-centimeter size. (Technically, this array was called super-WASP, and it was deployed in the northern hemisphere, with a copy located in the southern hemisphere.) The XO survey was a pair of similar cameras with a striking resemblance to the robot Johnny 5 from the movie *Short Circuit*. (The name "XO" is unique among astronomical surveys as it is neither an acronym, nor a reference to the XO telecommunications company.) The Hungarian-made Automated Telescope (HAT) network was an array of small telescopes spread across the globe to provide continuous observations—constantly dodging the daily motion of the Sun. And, to put a fine point on how small these systems are when compared to the "big glass" used for most astronomy, there was KELT—the Kilodegree Extremely Little Telescope.

With all these surveys running in the early 2000s, the field was poised to rake in huge numbers of new planet discoveries. Except that it wasn't. Finding a planetary transit is more than looking at a bunch of stars and seeing whether or not they get dimmer. Stars are messy objects. They pulsate, they have spots, they emit flares and other bursts of energy. Each of these processes causes the brightness of the star to change—potentially hiding or mimicking a signal from a planet. The Earth doesn't help things much either. Scintillation from the atmosphere, clouds, dust, city lights, and the Moon all affect the measured brightness of target stars.

The Earth's rotation alone caused problems. Most surveys took place at single sites, meaning they could only observe during the nighttime hours at that location. This left gaps in their data large enough to smuggle past a lot of undetected planets. Since most early discoveries were hot Jupiters with orbital periods near three days, we occasionally got really unlucky and looked at stars with planets whose orbits were almost an exact multiple of a day. If these planets transited during daylight hours, it could be years before the few-minute difference in its orbital period from a multiple of twenty-four hours would cause the transits to drift into nighttime hours—finally allowing it to show up in the data.

A more important issue that these surveys faced was noise from the atmosphere masquerading as planetary signals. There is always going to be some kind of noise in any detector. For these telescopes, the noise comes in part from the fact that the amount of light (the number of photons) arriving at the camera fluctuates slightly from one instant to the next. There is no way around this *shot noise*, because it is a fundamental property of quantum particles like photons of light. Outside the atmosphere, these fluctuations occur entirely at random. Because it is entirely random, having equal probability to fluctuate at any instant in time, it also gets the moniker *white noise*, similar to *white light* which is equally bright at all wavelengths.

The atmosphere, however, distorts the path that the light takes and causes the noise fluctuations to happen on different timescales—the timescales of passing atmospheric disturbances. A gust of high-altitude wind, or drifting wisps of water vapor, can cause the observed noise in one instant to become correlated with noise at another instant. If your target brightens slightly at one instant, it may be more likely than not to stay brighter in subsequent measurements. Similarly, if it has a slight dimming, say from light being deflected by a passing blob of air, it will be more likely to have neighboring observations that are also dim. Sadly, the timescale over which this atmospheric noise is correlated in this manner can be hours—comparable to the timescale of a planetary transit. This kind of correlated noise is called *red noise* because the timescale of the correlation in the noise is longer than the time between observations—referencing the fact that the red part of the rainbow corresponds to longer wavelengths of light. Successive observations will have brightness fluctuations that are more similar to each other. (Blue noise, for these purposes, would be noise fluctuations that are correlated on timescales shorter than the time between measurements.)

The existence of red noise in the data caused major headaches for subsequent observations of the potential planetary systems from our early, ground-based data. Even if there were no noise in the data, for every signal that you see that looks like a planetary transit, you would need to take a number of additional observations, using a variety of different instruments, to verify that the signal was caused by an actual planet. These observations were used to eliminate background stars, or spots, or small orbiting stars, as the explanation for the transits—removing potential sources of false positive signals. With red noise in the data, the number of false positive signals was quite high. Consequently, astronomers spent a lot of time observing stars where there were no planets to be found. There wasn't even any interesting astrophysics, like multistar systems or star spots, to serve as a silver lining to our cloud of dashed hopes. It was just noise.

We were spinning our wheels as often as not, and it took some time to sort out the best approaches for how to deal with all the false alarms.

A pair of conferences in the late 2000s, about a year and a half apart, illustrate the collective mood of exoplaneteers at this time—at least their mood from the point of view of a brand-new member of the group (me). In September 2006 there was a "Transiting Extra-solar Planets Workshop" held in Heidelberg, Germany.[3] By this time, most of the ground-based surveys were up and running, and a few planets had already appeared in the scientific literature. Kepler was still a few years away, but with all of the ground-based work, one had reason to believe there would be a large supply of new exoplanet discoveries.

However, at this workshop, it seemed like every other talk was about red noise. Statements to the effect of "We saw lots of candidates, but they turned out to be red noise" and "We were expecting to find planets, but the red noise kept getting in the way" were common. Each spurious, noise-induced signal was consuming a lot of telescope time. Follow-up observations that were intended to confirm a planetary transit signal were instead being spent chasing noise fluctuations.

Nevertheless, at the workshop there were also scientists working on solutions to this issue. One talk that stood out to me was from a scientist at Tel Aviv University, Tsevi Mazeh. (He would eventually be a future colleague on Kepler, though at the time I had no idea who he was.) He had developed a method for analyzing data from transit surveys that mitigated the effects of red noise to better extract real planet signals from those data. During his presentation, he offered to "clean" anyone's data for them to help make the discoveries that their instruments and efforts were designed to make. Nevertheless, while there were a few new results, the disappointment and frustration of my colleagues at this workshop was palpable.

Fast forward eighteen months to May 2008 at the 253rd Symposium of the International Astronomical Union in Cambridge, Massachusetts. By that time, many of the issues with noise had been addressed and new discoveries were pouring in—including science that went beyond merely the detection of new planets and into understanding some of

---

3. Standard spelling in the discipline wasn't yet established, hence the hyphen in "Extra-solar."

the details about them. That meeting had discussions about measuring the properties of exoplanet atmospheres, detecting light coming directly from the planets (thermal emission from the planet's internal heat), and mapping the planet's surface. It included several studies about the dynamics of planetary systems, and the discovery of strange planets where their orbits were wildly misaligned—sometimes even backward—relative to the spin of the host star.

At that meeting there were also talks about several space missions that were looking at exoplanet transits, including EPOXI, MOST, and CoRoT. To briefly introduce these missions, the CoRoT mission (for "Convection, Rotation and planetary Transits") was a European mission, designed to study stars by monitoring their brightness variations. It had a large field of view, so it could observe many stars simultaneously, lending it to finding new planets the same way that Kepler would. But planet detection was only part of the mission's science program. The remaining two missions were not survey missions for exoplanet discoveries, rather they were used to study known planets with better precision. MOST (which stands for "Microvariability and Oscillation of STars") is a Canadian satellite that looks like Spongebob Squarepants. It was originally designed to study stars by looking for small changes in brightness. As with CoRoT, exoplanet science with MOST was a secondary consideration and just came along for the ride. EPOXI was a two-part NASA mission—a combination of the Extrasolar Planet Observation and Characterization (EPOCh) and the Deep Impact eXtended Investigation (DIXI). The spacecraft itself was actually the repurposed satellite from the Deep Impact mission which had launched a large, half-ton projectile at a comet, smashing into it at over twenty-thousand miles per hour in order to measure the composition of the resulting cloud of debris. Once that mission was complete, the satellite (minus the projectile) was rechristened to look at exoplanets.

With new tools available, more planet discoveries, and more ways to study the systems, the Cambridge meeting was a watershed moment in the field. The results kept getting better and better as the week went on. Everyone attending knew that they had been part of something special. At the end of the week, during the concluding remarks, the organizers dressed like hippies from the late 1960s—comparing this conference to Woodstock. While all of this new territory was being explored by ground-based surveys and repurposed satellites, and discussions of these discoveries were filling the air whenever two exoplanet astronomers came within earshot of each other, NASA's Kepler mission was in its final stages of construction and testing. It was scheduled to launch in less than a year—looming large over the entire scene. In the next half decade, it would completely revolutionize our perceptions of these distant worlds.

Unlike everything else we've seen so far for the study of exoplanets, Kepler was a large mission designed explicitly with the purpose of finding planets—everything else Kepler might happen to discover was superfluous. A successful survey for transiting exoplanets requires a large sample of target stars to overcome the rarity of individual transit events, continuous coverage so that you don't miss any transits that do occur, high-precision sensitivity to light so that you can find small planets, and low noise so that the planet transits don't become obfuscated by random variations in the electronics or spurious signals. These considerations drove the design of the Kepler mission. Jon Jenkins, a scientist at NASA Ames and a central figure on the Kepler team, once compared it to a thoroughbred racehorse—it was built for one reason, to find planets, and it would be good at it. [5, 16].

The most important part of the satellite was the 95-megapixel digital camera, or *photometer*. The camera was the size of a small cookie sheet and was made from forty-two rectangular CCD chips (2200×1024 pixels each). The array of chips was so large that the camera surface had to be curved so that all parts of the camera would simultaneously be in focus. The chips were laid out in a pattern with a specific rotational symmetry. When the camera is rotated about its center point 90 degrees, all of the gaps between the chips line up in the same place. (The only exception to this was the pair of chips in the middle, where they lined up after a rotation of 180 degrees.) This fourfold symmetry of the camera footprint allowed its field of view to remain roughly constant as the satellite turned each quarter so that the spacecraft's solar panels remained facing the Sun. Plate 3 shows the first image that Kepler took of its field of view, its "first-light" image, where you can see this symmetry in the camera layout. Plate 4 shows the final image taken by the Kepler telescope, which shows that portions of the photometer degraded over time.

This photometer was mounted inside the telescope, pointing back toward the primary mirror in the telescope base, and situated where the mirror would focus the light. Kepler is a *Schmidt* telescope—a designation that, when I heard it mentioned, seemed like it should have filled me with wonder and appreciation, but I could never say that it did. What I can say is that this design allows for a large field of view. In the language of telescope designers, Schmidt telescopes can be very "fast," meaning that they can focus light from a wide angle onto the detector, while simultaneously having a relatively short length. Because of the wide angle of incoming light, the images from Schmidt telescopes are distorted and need a correcting lens placed at the top of the telescope to remove those distortions. For Kepler, this meant that the telescope aperture at the correcting lens (95 cm) was a fair amount smaller than the mirror diameter (1.4 m).

The upshot of Kepler's design was that it could image over 100 square degrees on the sky (essentially 10 degrees on a side). That patch of sky is roughly the size of the palm of your hand when held at arm's length. That may not sound impressive for a telescope, but let's compare it to the Hubble Space Telescope. The Hubble Space Telescope's most recent Wide Field Camera, WFC3, has a field of view that is 0.05 degrees on a side. If you were to hold a penny at arm's length, 0.05 degrees is the apparent size of Abraham Lincoln's eye. Hubble's *wide field* camera is fifty thousand times smaller than the patch of sky that Kepler can observe. By the standards of astronomy, Kepler's field of view is enormous.

Given this telescope and camera, we still need to know where to point it and for how long. The primary design criterion for the mission was to detect a distant Earth-like planet, in an Earth-like orbit, around a Sun-like star. But just seeing a single transit wouldn't be enough. We needed to measure the planet's orbital period, which means we needed at least two transits. And we needed to know that the orbital period we measured was, in fact, correct. These requirements mean that, in order to claim a planet discovery, the Kepler team needed to measure at least three transits of a given planet, since having three transits allows us to make two separate measurements of the orbital period (once between the first and second transits, and again between the second and third). For a one-year orbit, the minimum length of our observations is two years and a day. However, that assumes that we see the first transit the day we turn the camera on. The chance of that happening is 1 in 365. It's equally likely that we would be just as unlucky, and had one of the planets transit the day before we started observing. So the mission strategy was to continuously monitor the field of stars for four years. (Eventually, to save some money, the mission was shortened to three and a half years.)

The requirement for four years of continuous observation provides at least one constraint on where to point. It can't point anywhere along the plane of the Earth's orbit. If it did, then the Sun would eventually move into the field of view, blocking the light from the target stars and destroying the instrument in the process. Kepler needed to point either above or below the Earth's orbit—narrowing the options slightly. Additional considerations for choosing the field of view came down to a balancing act between competing issues. You need to look at a part of the sky with enough stars to conduct a thorough survey. The most crowded collection of stars in the sky would be the plane of the Milky Way itself. But, if you have too many stars, you reach the *confusion limit*, where the light from multiple stars blends together so that you can't distinguish between them.

Reaching this confusion limit causes a number of problems. For example, interloping stars that happen to align with your target star will dilute a planet signal. A transiting planet only blocks a small fraction of the light from one of the stars, while the interlopers shine uninhibited—making the observed transit signal smaller than it would otherwise be. For any planet that you still detect, you would infer the wrong planet size. It would appear smaller than it is in real life because it isn't blocking light from the other stars that are blended with the target star. It may block one percent of the light from the star that it orbits, but it may only block a half percent of the total light from all the stars.

Beyond the confusion limit you also run the risk of mistaking a signal as planetary when, in fact, it comes from some other source. For example, if one of the stars that is blended into your image has spots, the star spots rotating into and out of the image could look like a transit signal. Or you may have an eclipsing binary star system that is blended into the target star. Here, the eclipses of the intruding binary system would mimic a planet signal. These consequences of confusion mean that you can't operate the mission if the number of stars in the field of view is too large.

The confusion limit of Kepler was exacerbated by the fact that the pixels on the camera were huge. Comparing again with WFC3 on Hubble, each side of a pixel on Kepler is a hundred times larger than one on Hubble, and the pixel covers ten thousand times the area. Hubble can resolve a detail that is a hundred times smaller than what Kepler can see. So a collection of stars that Hubble would easily identify as separate entities would turn into a single bright pixel on Kepler. Kepler's fat pixels also mean that the image quality is relatively poor, especially when compared with Hubble—its pictures are blocky and lack definition. But Kepler's "camera" was a photometer, not an imager. It was never meant to take pretty pictures, it was meant to measure brightness— to count photons. As with virtually every aspect of the mission, the camera design was a trade-off between competing demands, and the overarching demand was planet finding.

While the confusion limit with Kepler forces you out of the plane of the Milky Way, if you go too far the stars become so spread apart that you can't get the necessary numbers. In the end, the science team chose a field of view slightly to the north of the Milky Way, midway between the constellations of Cygnus and Lyra. One corner of the camera juts down into the galactic plane, and the opposite corner projects high above the plane. When you look at a full image from Kepler, you can see how the stars are more concentrated as you move diagonally across the field toward the Milky Way.

Ultimately, the number of stars that were surveyed by the mission is roughly 150,000, which varied slightly as some stars near the edges of the chips could be missed from one quarter to the next. The final decision on where to point the telescope was dictated by the positions of the brightest stars in the sky. Really bright stars would saturate the camera, spoiling the measurements of the dimmer stars in their vicinity. To prevent this from happening, the camera was oriented so that the handful of particularly bright stars would consistently land in the gaps between neighboring CCD chips as the spacecraft rotated every quarter.

The number of target stars wasn't chosen because it was the complete list of suitable targets in the field of view—there were millions of stars that Kepler could see. Instead, the size of the survey was limited by the bandwidth needed to beam the data back to Earth. Communicating with Kepler required the use of NASA's Deep Space Network. This is a network of radio antennas spread around the globe—one of the world's most expensive internet service providers, since their coverage includes the entire solar system. Unlike many satellites, the Kepler spacecraft doesn't orbit the Earth, it orbits the Sun. This orbit was chosen because circling the Earth induces both temperature variations and noise from the Earth's magnetic field, both of which affect the instrument sensitivity. Getting the spacecraft away from the Earth meant a Sun-centered orbit that is slightly longer than a year, and distances between the Earth and the satellite that are quite large. Hubble is only 350 miles from the Earth's surface. Kepler passed the 350-mile mark about an hour after it launched. Because of the large distances between Kepler and Earth, we needed this more valuable communication system.

The limited bandwidth available for data transfer from the spacecraft was the fundamental bound on the number of targets selected for observation. Or, more accurately, it set the bound on the number of pixels used for the observations. To save space, both on the onboard memory and when downlinking the data, Kepler would only store a portion of the images that it took, rather than storing its entire view of the sky—most of which was dark anyway. Only specific pixels surrounding each target star were retained on board and transmitted back to Earth. These "postage stamps" meant that Kepler could observe for a longer period of time before its memory filled, and before it turned its antenna toward the Earth for downlink. It would spend less time yammering with the scientists on the ground, and more time finding planets. (Originally, the spacecraft was designed with a steerable antenna so that it could continue its observations while communicating with Earth, but as the price tag mounted, that was a feature that was scrapped. On a spacecraft, moving parts mean risk, and risk means money.)

In the end, Kepler would take and store a single full-frame image each month as a means to monitor the overall health of the camera. For the rest of the observations, it stored only the pixels assigned to the target stars, and a few pixels used for calibration. The number of pixels assigned to each target depended on how bright that star was— brighter stars required more pixels. Despite the brightest stars being deliberately placed in the gaps between CCDs, many of the Kepler targets were still bright enough to saturate large collections of pixels. The way CCDs work is that an incoming photon dislodges an electron, which is physically trapped in the structure of the semiconducting pixel. If too many electrons are dislodged, they spill from one pixel into another, like flowing water filling an ice-cube tray. The images of those stars appeared as long streaks on the camera—distorting what the stars actually looked like.

Since the mission is fundamentally about measuring brightness, the fact that the signal spilled into other pixels wasn't necessarily a problem. Each electron represented a photon, so as long as the electrons weren't lost, the important information was still there—you just needed to assign the spilled electrons to the correct target. This meant that the postage stamps for the brightest targets were large, weird-shaped sets of pixels, while dimmer targets would just be those pixels that immediately surrounded the intended star.

The way that the pixels were assigned to the different Kepler targets is an interesting, technical issue. Kepler could have no more than 170,000 targets, and no more than 20,000 of them could exist on any one chip (or, no more than 10,000 targets for each half of one chip). The average number of pixels assigned to a target star was about 30. After some time operating the camera, the science team decided it would be wise to include a sample of different parts of the Kepler field that didn't have any stars in them. This gave us a means to monitor background noise, both from the camera and from brightness variations of stars that were otherwise too dim to resolve. Ultimately, there were nearly 1000 calibration pixels selected for each chip, which were generally divided into 3×3 *super-pixels*. Between the target stars and these super-pixels, only data for 5.5 million pixels, out of the 95 million on the camera, were stored on the spacecraft and downlinked to Earth—a little less than six percent.

Kepler made its observations on two different exposure timescales, or *cadences*. Most observations were made with the *long cadence* of once every half hour. There was a special allocation of 512 targets that were observed with the *short cadence* of once every minute—with only a maximum of 43,520 pixels to share among them. Each Kepler "observation" was itself a set of steps that was more complex than one might initially imagine. You don't just point the camera and hold down a button for the desired amount of time. If you did that, the electrons dislodged by the light from the brightest targets would flood large swaths of the camera, ruining the observations of the other targets. Instead, the camera collected data for six seconds (6.02 seconds to be exact) before the CCD chips were read out (meaning that the dislodged electrons in each pixel were counted), and the data stored. The camera readout itself takes a half second (0.52 seconds). Since Kepler has no shutter, the entire time that the camera is being read out, more light is hitting the photometer, smearing out the image of each target during the readout process. This smearing also must be accounted for in the analysis.

This series of steps repeats itself, with each new six-second observation being added to the existing data until the desired length of the cadence is reached for each target. For the short-cadence data, it takes nine cycles to complete the observations—giving a total time interval of 58.84876 seconds. (I'm being annoying with these numbers here for a reason; bear with me.) To get data for the half-hour, long cadence you continue aggregating observations for thirty more cycles (of nine sets of six-second observations), which gives a total time interval of 29.4244 minutes. At that point, the data are stored on the spacecraft, and it starts its observations for the next cadence.

This degree of precision for the timing of these steps appears to be overkill, and for most applications it probably is, but the computer on Kepler needs to know exactly what it is supposed to do and when. You can't just tell it to "take pictures every so often and get back to me." Each cadence is timed to within a tiny fraction of a second, and synced with an imaginary clock at the center of the solar system, and with the clocks on Earth, while both the Earth and the Kepler spacecraft are constantly drifting relative to each other (and the solar-system center) because of the small eccentricities of their orbits. Without this careful synchronization, the time series of the observations wouldn't be accurate, and a fair amount of the science that we wanted to do with the data would be compromised.

Once per month the data are downlinked through the Deep Space Network to the ground, where they are processed. Here too it isn't really a month: Kepler's orbit is about 373 days (just over fifty-three weeks), and the data are downlinked twelve times per year. With every third download, the spacecraft rotates 90 degrees to reorient the power-providing solar panels toward the Sun. Now, with the data on the ground, we can start considering the idea of looking for planets, or at least we can think about beginning to start considering the idea.

The raw data are a mess. Having the noble goal of finding planets in mind, and then seeing what the data look like when they arrive on the ground can be disheartening. The desire to detect an Earth-sized planet on a one-year orbit around a Sun-like star yields the design criteria for the mission's photometric sensitivity. Kepler needs to detect changes in brightness of only twenty parts per million on a bright target star over the course of six and a half hours (half of the time needed for the Earth to transit the Sun). A precision that small means that everything is a source of noise. Every change in the temperature of the spacecraft can be seen in the data, sudden drops in the sensitivity of a pixel are common, as well as electronic noise in one wire when it senses the electrons flowing in a neighboring wire (crosstalk). Some CCD chips have a band of noise that periodically creeps across them. Cosmic rays impact the camera, leaving bright spots in an exposure. Occasionally, huge parts of the camera would light up in what was called an *Argabrightening* event—named after the Ball Aerospace engineer Argabright who discovered them. (Our best guess for what caused these were dust particles, or chips of paint, that were flaked off the spacecraft due to micrometeorite impacts. The particles would drift across the telescope aperture and scatter sunlight into the camera.)

Another source of noise that becomes important when we are talking about part-per-million measurements is the fact that the pixels of the camera themselves are not uniformly sensitive. That is, a single pixel can have one corner with one sensitivity and another corner with a different sensitivity. Small changes in the orientation, or *attitude*, of the spacecraft can place a star on a slightly more or slightly less sensitive part of that pixel. The difference in sensitivity across the pixel would look like the changing of the brightness of the target star—thereby mimicking a planetary signal. It doesn't take much change in the attitude of the spacecraft to look like a ten-part-per-million brightness change—a drift of less than one ten-thousandth of a degree is enough.

In fact, the non-uniform sensitivity of the pixels, and the limits to which we could point the spacecraft, were sufficiently important that the science team considered deliberately defocusing the images. The idea was that defocused images would spread the light more uniformly across several pixels, so that changes in the measured brightness that would result from small changes in where the image lands on the camera would be washed out by the fact that the light is already spread across a larger number of pixels—averaging over the sensitivity variations. This technique had been used in other contexts, but in the end we didn't adopt it for Kepler. Instead, we focused the light as best we could across as much of the camera as we could. This meant pointing the telescope as consistently, and as stably, as possible. Nevertheless, small temperature variations that occurred as different parts of the telescope warmed throughout its orbit caused the focus to change, the mirror to warp, and the images to drift in the camera. Even the tiny *velocity aberration* that occurs because of the relative motion of the spacecraft and the distant stars caused the images to drift in the camera by a half pixel.

All of these changes induced brightness variations that could hide the planetary signals we wanted to see.

To correct for all of these noise sources, we fed the data through a piece of computer software called the data analysis *pipeline*. Initially, this pipeline ran on a dedicated cluster of computers with seven hundred processors running day and night. Data processing went fairly quickly at the beginning because there weren't many data to examine. But, with each new downlink, the amount of data grew. To get the best results, all of the data for each star were completely reanalyzed with all of the additional data being combined with the previous data in order to look for both smaller signals and signals from planets with longer orbital periods. Eventually, it took longer to analyze the data than the time between downlinks. The analysis of the first two years of data took ten months to complete—and new data were arriving every month. Consequently, in 2011, the pipeline software was installed on the gigantic Pleiades computer located at NASA Ames. At the time, this grocery-store-sized computer, with its fifty thousand cores, was the fourth most powerful computer in the world. Now, more than a decade later, it is still in the top hundred.

The analysis pipeline would churn through the data, removing the effects of noise, removing ghost images (caused by reflections within the telescope), and eventually removing brightness variations that came from the stars themselves—star spots, pulsations, and rotations—to leave behind an otherwise pristine, constant brightness for the target stars, that is, constant brightness except for the small dips that came from the transits of the orbiting planets. In many instances, the transit signals of the discovered planets were hundreds of times smaller than the brightness variations caused by different instrumental effects, noise sources, and stellar variability.

Each time it ran, our analysis produced a catalog of *threshold crossing events*, or signals that looked promising as planetary candidates. At first, each of these events was examined by eye to help separate the good candidates from false positives. But the plan from the start was to completely automate the process. By the end of the mission, with continual improvements to the analysis software, the data processing, quality diagnostics, and identification of planet candidates were turned over to the machines.

A lot of the signals were obviously good planets, but the smaller the signal, the more challenging it becomes to say so with confidence. And it is the smallest signals that are the most interesting from the mission perspective. Thoroughly combing through the list of threshold crossing events produced the penultimate catalog of *Kepler Objects of Interest* or *KOIs*. It was this list of KOIs that the Kepler Science Team used to organize the bulk of their follow-up work. We would examine the objects of interest with other telescopes and additional software tools to hopefully promote them from objects of interest to *planet candidates*, and eventually (often several years later) to verified planets. For most of the mission, it was the KOI list and the planet candidate list that were the spring from which other science flowed. These were the lists that would be distributed to the science team and to the broader community to be analyzed, observed, studied, modeled, and discussed.

3

Early Results and the Science Team

As the Kepler mission developed, it continually needed new expertise and new information to prepare for its operations. Aside from ensuring that the satellite functioned properly and could meet its design sensitivity, it was crucial to know which stars would be on its list of targets. As we've seen, it isn't like you can just point it to a place in the sky and hope for the best. Data storage on the spacecraft was limited, as was the link to receive information on the ground. You can't waste those resources just because of poor planning. In the years leading up to launch, the stars in the Kepler field of view were studied by a large, ground-based, observing campaign to produce the Kepler Input Catalog (KIC or "kick"). It was from this catalog that the team selected the targets most amenable to finding planets—especially the Earth-like, habitable-zone planets that were central to the mission, the *habitable zone* being the region in a planetary system where liquid water (and hence life) can exist.

An important issue to consider for target selection is that stars go through several phases in their lifetimes. In particular, stars near the end of their lives go through dramatic changes in their size and brightness. For example, while the Sun will remain largely as it is today for the next five billion years, in its last hundred million years it will grow in size by about a factor of a hundred, shedding a bunch of material along the way, as it becomes a red giant star. This causes problems for planet detection because, while the star grows, the planets stay the same size. Assuming that the planets survive the star's expansion and mass loss, if you originally needed twenty-parts-per-million sensitivity to find a planet around that star, you now need sensitivity that is ten thousand times better—a few parts per billion.

All stars go through this late phase in their life, but only a small fraction actually experience it at a given point in time. This giant phase only occurs during the final few percent of a star's lifetime. That fact would seem to work in our favor since, all other things being equal, only a small fraction of the target stars would be giants. However, with that increase in the size of the star comes a thousandfold increase in brightness. Larger, more luminous stars can be seen from farther away. For stars that are a thousand times brighter, the volume of space where they can be seen increases by a factor of ten—counteracting the relative rarity of this type of star. Any survey that is limited by the brightness of the target stars will automatically have an overabundance (but not necessarily a majority) of brighter, yet more distant, giant stars.

When astronomers talk about their surveys, the two main types that they refer to are *magnitude limited*, which are easier to make but suffer from this bias, and *volume limited*, which are harder to conduct but which examine all candidates within a specified volume and don't have this bias. Kepler, fundamentally, was a magnitude-limited survey, so the team had to address the problem of distant, bright stars contaminating the target list. Resolving this issue was one of the primary purposes of the Kepler Input Catalog.

That catalog included observations of over thirteen million stars. The inferred stellar properties from that survey were used to select dwarf stars (as in not-giant stars), with a mass roughly equal to the Sun, and which were unlikely to be in multistar systems. Ultimately, only about half of the stars in the KIC survey landed on the footprint of the Kepler photometer. The fact that we reduced the list of stars from the five or six million initial possibilities down to 150,000 shows how picky the mission requirements needed to be—only the best three percent of the possible stars made the final target list.

The KIC survey was run by a handful of scientists, notably David Latham and Gilbert Esquerdo at Harvard, Timothy Brown at the Las Cumbres Observatory, and Mark Everett at the National Optical Astronomy Observatories [17]. The camera on their telescope didn't cover as much of the sky as the Kepler mission could see, so the Kepler field was surveyed by stitching together sixteen hundred smaller, overlapping images taken with their camera. (They actually used a few different cameras throughout the survey, but the so-called KeplerCam was the main workhorse.) This survey used a variety of filters on the camera, which only allowed certain colors of light to be recorded. By comparing the brightness of the stars in these different colors, the astronomers could determine some of the star's essential characteristics—the most important being the temperature and the strength of the gravitational field at the stellar surface. These two quantities are a good diagnostic for separating main-sequence, Sun-like stars from giant stars because, when a Sun-like star passes into its red-giant phase, it becomes cooler and larger in size—it's fluffier. The larger the star becomes, the weaker the strength of gravity at its surface. So, by eliminating cool, low-gravity stars from your list, you eliminate the contaminating giants.

For months prior to the mission, members of the science team whittled down the list of potential targets to get to the final set. As those preparations drew to a close, a sample of smaller stars was added to the mix—M-dwarfs. Stars are classified into seven broad categories (O, B, A, F, G, K, and M) depending primarily upon their mass. The Sun is right in the middle of the G-type stars. Kepler's primary targets, intended to be Sun-like, focused on G and the neighboring F and K stars. F-stars are slightly larger than the Sun, and K's are slightly smaller. M-dwarfs are a class of stars that are significantly smaller, and much dimmer than the Sun. Despite their small size, these stars are plentiful—outnumbering the combined F, G, and K stars by more than three to one. The dimness of M-dwarfs would normally make them poor candidates for Kepler, since the brighter the star, the more precisely we can measure that brightness. M-dwarfs can be a thousand or even ten thousand times dimmer than the Sun.

However, these stars are also small—roughly the size of Jupiter. Because they are so small, planetary transits block a larger fraction of their light. These prominent transit signals counteract the dimness of the star, and facilitate the detection of their orbiting planets. Moreover, since finding exoplanets in a star's habitable zone was a primary goal of Kepler, M-dwarfs presented an important opportunity. Because they are so dim, their habitable zone is much closer to the star, where planetary orbits are shorter. Rather than transiting once per year, a planet in the habitable zone of an M-dwarf would transit every few days. These benefits motivated the science team to include some five and a half thousand M-dwarfs in the Kepler target list that might otherwise have been left out.

Even with the careful selection of high-quality targets, several obstacles still prevented the science team from declaring new planets with each transit signal in the Kepler data. Binary stars cause all sorts of problems for planet detection, especially those that eclipse each other. The eclipses produce a signal similar to that of a transiting planet. Often, stellar eclipses are too deep to be confused with planets. If the stars are equal sizes, the light can drop by half. However, if the stars appear to only graze each other, clipping just a small fraction of the light, the change in brightness is similar to a planet transit. Or, if a given target has an interloping eclipsing binary in the background, the eclipses of the background binary may only affect a small portion of the total light—mimicking a planetary signal. (For example, if the background eclipsing binary contributes two percent of the total light, and the eclipses block half of the light from the binary, you get a one-percent drop in brightness, just like a transiting Jupiter-sized planet.) These kinds of signals aren't noise—they come from real astrophysical systems, but they aren't planets.

Despite our best efforts to select isolated stars, some of the target stars were still members of binary star systems, or they were targets whose light was blended with binary stars coincidentally aligned along the line of sight. To disentangle these sources of error, a number of astronomers were brought onto the science team to conduct high-resolution imaging, including adaptive optics imaging, of potential planet-hosting stars. High-resolution images have less blur around a given star, and can resolve the presence of additional stars that would otherwise be invisible to Kepler with its poor image quality. These observations meant we could account for stray light from interlopers in our analysis. By treating these scenarios properly, planets orbiting blended targets weren't immediately tossed out.

Weeding out astrophysical false positives, like eclipsing binary stars, requires other ground-based observations besides imaging—this time using spectrographs. Several team members were tasked with follow-up observations to search for chemical signatures of multiple stars. This was especially important when trying to identify stars that were too close together to resolve with imaging. The chemicals that show up in a star's atmosphere depend upon the temperature of the atmosphere. Higher temperatures excite higher energy states, while lower temperatures excite lower energy states. Different elements will have a different mixture of states with different energies.

Some of these spectrographs could also look for a Doppler-shift signature. If the object that caused a transit signal was actually a star, the associated Doppler signal would be hundreds or thousands of times larger than what a planet could produce—marking the signal as a companion star. Observations like these, which are primarily made to eliminate the possibility of mistaking an eclipsing binary star for a planet, bring the added bonus of improved measurements of the properties of the target star beyond what we get from the Kepler Input Catalog. The input catalog only looked at the stellar brightness in a handful of broad swaths of the electromagnetic spectrum. The *reconnaissance* spectroscopy measurements, designed to catch masquerading binary stars, weren't good enough to positively identify planets, but they provided important details about the stars that would give more insight into any planet that was discovered—a better understanding of the star means a better understanding of the planets orbiting that star.

For a transit signal to be an actual discovery, it still had more hoops to pass through. Not only did the team verify that the signal was not due to noise, a *false alarm*, they also ensured that it wasn't from some multistellar system, a *false positive*. Reaching this stage brought the object of interest to the exalted level of a *planet candidate*. This is where a large fraction of potentially planetary signals remain. Thousands of discoveries are still languishing in this state of maybe-a-planet-but-maybe-not. (As of this writing, almost a decade after the Kepler mission finished, about half of the relevant signals remain planet candidates.)

The list of planet candidates is still useful for studying the properties of exoplanets in general, even if some of them are not actual planets. If you are doing a statistical analysis that examines the properties of a group of planets, like estimating the number of Earth-sized planets in one-year orbits, having false positives in your catalog can be OK as long as you have a way to account for the frequency with which spurious signals appear in the data. You might not know which planets are real and which are not, but you can know how many there are of each.

Studying groups of planets may leave some wiggle room for error in how the planets are classified, but if we want to claim that a given candidate is a real planet, and to understand the properties of that planet, we need to be sure that it is not a false positive. Once Kepler got going, each dip in brightness of a target star was vetted with increasing scrutiny in order to elevate it to the status of a confirmed planet. Initially, the only way to do this was the old-fashioned way, measuring the Doppler shift of the planet as it orbits the host star. Confirming planets with high-precision Doppler measurements, combined with the high-resolution imaging and the reconnaissance spectroscopy of the host star, was the bulk of the work done by the Kepler Follow-up Observing Program. Several members of the Kepler Science Team were devoted to conducting these follow-up measurements using telescopes spread across the globe—Arizona, California, Texas, Hawaii, and the Canary Islands.

In the two decades between the initial mission concept and the mission launch, most of the effort went into the essential aspects of getting Kepler off the ground and producing a catalog of results. Beyond the scientists and engineers working on the spacecraft, the data analysis pipeline, and the follow-up observations, the science team included a few people whose task it was to make sense of it all. These included theorists who build computer models of planet formation, or who study the evolution of planetary orbits, or who investigate the properties of planets themselves. While Kepler was being designed and built, these scientists helped make predictions about what we could expect to find in the Kepler data.

Prior to launch, you only needed a few people to prognosticate on the broad implications of what Kepler would discover—you needed more people whose job was to bolt the satellite together. However, there is a lot of science to be done with Kepler data that goes beyond these initial predictions. Once the mission launched and the data started pouring in, it needed new expertise to fully understand the implications of the discoveries. To expand the scope of the scientific footprint of the mission, and to add warm bodies who could help sift through all of the data, NASA announced the Kepler Participating Scientist Program in February 2007—less than one year after I graduated from the University of Washington [18].

Participating scientist programs are common on NASA missions because of this change in their needs as they evolve from planning, to construction, and eventually to operations. At the time of this request for Kepler, there were thirty members of the science team, not including the operations and mission management, and there were five broad areas of expertise, which were transit follow-up, analysis pipeline development, precise target position measurements (astrometry), the characterization of planetary systems, and asteroseismology (or the study of how stars shake). This last topic, oddly enough, comprised the largest group of people expecting to look at the Kepler data as they came from the spacecraft. The asteroseismology group was a huge international collaboration with hundreds of scientists, two of whom were also members of the Kepler Science Team. They were all waiting to look at the data from the stars—especially those that did not have planets orbiting them. (It may come as a surprise to anyone who attends an astronomy conference today to realize that there are areas of astronomy aside from exoplanets and cosmology. Many astronomers actually study stars.)

In order to join Kepler as a participating scientist, candidates submitted proposals to NASA that outlined what science they would do for the mission. Then, after a review, NASA selected from among the proposals those who would be brought on board. NASA initially added eight new people to the team, including me, with a ninth member brought on shortly thereafter. Most of us who joined were looking at planet properties or planetary system properties, with a couple of us having plans for follow-up observations and making improvements to noise mitigation in our analysis. We all arrived on the scene roughly one year before the newly rescheduled launch of November 2008, which had just been pushed back from June 2008.

The first meeting that I attended with the expanded science team took place in Boston. I knew a few of the team members by name, but only a couple personally. Most attendees at this meeting were completely new faces to me. Or, in one case, the shadowy silhouette of a face. Among the most memorable experiences from that meeting was seeing Natalie Batalha speak. Natalie, a future leader of the science team, was participating online, and the combination of lighting and the 2008 technology produced a talking shadow that would make any FBI informant proud.

The early team meetings were devoted to topics that pertained primarily to the launch. The standard fare was subjects like the final tests of the spacecraft instruments, the discoveries of sources of noise and how they would affect the mission, completing the development of the data analysis pipeline, target selection, and coordinating the efforts for the ground-based, follow-up measurements. For me, the first couple of meetings were largely trying to figure out what was going on, who all my new colleagues were, and what role they played on the mission.

At this Boston meeting, we participating scientists introduced ourselves and gave short presentations about the work we would do. My presentation was, apparently, not very memorable since the only reason I recall these presentations happening at all was that one of them was about measuring precise transit times—which was the data I would need for my work. It took these first meetings, and the intervening months, for us to see how our scientific interests complemented and occasionally overlapped each other, since several of the new members had similar or related projects in mind.

As is typical with NASA missions, the launch date slipped again, now scheduled for March 2009. This date was sufficiently serious that the team meeting was slated to take place in Florida to coincide with it. As we were preparing to travel to this ultimate meeting, disaster struck NASA and caused major concerns for Kepler and for us scientists and engineers. On February 24, 2009, two weeks before the Kepler launch, NASA's Orbiting Carbon Observatory malfunctioned shortly after lift off. The *fairing*, the capsule that houses the satellite on the tip of the rocket, failed to separate from the rocket. The extra weight kept it from reaching orbit, and it crashed to the Earth. In light of this disaster, there was a scramble to ensure that Kepler's fairing wouldn't have that same problem [19].

Later studies of the crash of the Orbiting Carbon Observatory, after a second mission failed two years later from the same type of rocket, found that an aluminum supplier had been falsifying the certifications of their product. Fairings often separate from the rocket body using bolts that are blown by explosives, and since those explosions happen next to a few hundred million dollars worth of sensitive equipment, you want to use as little pyrotechnics as possible. So it is important that the material properties of the bolts you are trying to break actually match the specifications. Needless to say, NASA switched suppliers.

That was years into the future. For Kepler, we were using a different launch vehicle, but the crash warranted a reexamination of the Kepler system so that we wouldn't be left with a field of debris to scavenge for spare parts. It is difficult, after all, to salvage anything from a spacecraft after a high-velocity, uncontrolled reentry like the one experienced by the Orbiting Carbon Observatory. As the confidence in the Kepler delivery system mounted, the science team members made their way to Cape Canaveral to meet and to watch the launch. The excitement was tangible. The NASA veterans told us that its nighttime launch would be pretty spectacular.

Prior to launch, the team was briefed on the procedure. The satellite was situated atop a Boeing Delta II rocket, a launch vehicle with over a hundred successful flights. There were nine solid rocket boosters surrounding the main, three-stage, central cylinder. We were told that the mission would lift off heading south over the Caribbean, shortly thereafter six of the nine boosters would fall off, the remaining boosters would turn on, and the first stage would take the spacecraft out of the atmosphere. The remaining boosters would then come off, the first stage would detach, and the second stage would begin a brief burn. At that point, the fairing would separate, exposing the satellite, and the entire assembly would start rotating to keep it stably pointing in the right direction. The rocket with the Kepler satellite sitting exposed on the tip, would then drift around the Earth until it could start communicating with tracking stations in Australia.

Here, the second stage would reignite and burn until Kepler was sufficiently deep into space that the second stage could be discarded and the third stage engaged to inject Kepler into its final orbit. Once there, the satellite would still be spinning rapidly and hurtling through space, leaving the Earth and its gravitational field behind as it began its orbit around the Sun. To slow the spin, a pair of heavy, metal balls had been attached to the spacecraft base. The two balls were tethered to the satellite with lines that would extend to either side. As the cords attached to the weights extended from the satellite, the rotation slowed, just like when ice skaters slow their spin by extending their arms or legs. Once the balls reached the end of their lines, the tethers were cut, the balls were flung out into space, and the spacecraft was left behind with its much slower spin rate.

As this launch sequence was being explained to the team, I thought, "Wow. Rocket science really is hard."

Following our briefing, on the evening of March 6, 2009, the science team gathered on Florida's Atlantic shore to watch the launch. The anticipation rose, as the countdown worked its way to zero. There were two three-minute launch windows available to put the spacecraft where it needed to go. Liftoff was slated for 10:49:57 p.m. Eastern Time, while the second window was for 11:17 p.m. When the time came, a burst of light sprang from the ground as the rocket began its ascent. After rising what appeared to be only a few feet off the ground, it jettisoned six of the nine solid rocket boosters. It looked a little early to me since it seemed like the rocket had barely cleared the launchpad. No one else looked bothered—it did happen at the correct, one-minute mark, and we were watching from quite a distance from the pad, but it still looked like it hadn't gotten very high. The bright streak of fire and smoke continued to lengthen as the rocket arched to the southeast, eventually fading into the distance. Cheers and congratulations went all around as we received periodic updates of the spacecraft position and trajectory. Then it was over. The launch was a success, the Kepler spacecraft began preparations to take its first images, and we dispersed back to our hotel rooms.

There was more to this Florida meeting than just the launch. This was the meeting where our scientific collaboration really materialized. Jack Lissauer, a NASA scientist and long-time member of the Kepler team, was interested in the dynamics and properties of planetary systems. He invited several of the new participating scientists to a meeting over dinner—myself included. Since our scientific interests all overlapped, he wanted to hash out some of the details of how we would work together in the coming years. As we pulled tables together from around the restaurant, we had a chance to see who our closest colleagues would be.

Meetings like this, with people trying to carve out a space for themselves and their work, always have some element of awkwardness. Who will lead this study? Who will help with that project? How do we reconcile if multiple people want to lead the same thing? There are elements of both collaboration and competition. The challenge is striking a healthy balance, which takes good leadership and an open and collegial environment. This semicasual meeting, with a half-dozen people sitting at a few tables in a Florida restaurant, formed one of the great scientific collaborations of NASA's history, in my humble and totally unbiased opinion. In the coming years, the nascent Multibody Working Group would spearhead some of the most iconic results from the Kepler mission, and would influence the overall culture of the broader science team. For the moment, however, we were just getting our footing while simultaneously trying to avoid stepping on each other's toes.[1]

The spacecraft settled into its Earth-trailing orbit over the subsequent weeks. After a month, NASA deemed it safe to jettison the dust cover—Kepler's final layer of protection against the elements, and the final impediment to collecting light from distant stars. A one-minute exposure, taken on April 8, 2009, would be Kepler's "first light." This gigantic image with millions of visible stars, interspersed between the gaps in the array of CCD chips, gave the literal appearance of a window into the galaxy. This image is shown in plate 3.

Kepler was still in the commissioning phase, with data for several engineering and calibration measurements to be gathered before scientific observations could commence. One of the first, big operational questions posed to us following the launch appeared in an email that was broadcast to the science team. The first images taken by the spacecraft were out of focus. Although the camera functioned, and we would be able to detect planets, continuing operations in this way would add noise to the data that would limit our sensitivity to the smallest planets that Kepler was designed to see. The email essentially said, "The camera appears to work, but we'll lose a bit of our science because it's blurry. Should we focus the image? Or should we keep it where it is?"

We anticipated that we would need to change the focus before we started operations, but we still wanted to avoid doing anything that could hurt the spacecraft. Despite it being part of the plan, twisting that knob meant moving a part, and moving something is always a risky business when there is no going back. For Kepler, focusing meant moving the entire mirror with the camera remaining fixed. If the mirror got stuck, or didn't stop, the whole mission would be compromised. If we were to keep it as it was, the mission scientists estimated about a

---

1. We actually formed two working groups: the Multibody Working Group and the Transit Timing Variations Working Group. But they had the same members and were interested in essentially the same science. For a while we went with the Multibody/TTV Working Group. Then we just stopped caring about the distinction. The email list serve for the group used "TTV"; in conversation we were the "Multibody Group."

twenty-percent loss in output. So the question posed to the science team was essentially "Do we risk it all for that twenty percent?"

I kept my mouth shut. This decision was too big for me.

After an hour or so, responses to the question began trickling in. They were all variations on the same theme. The consensus was broadly that "we trust the engineers." We hadn't come this far, and worked for so long, just to hamstring the spacecraft because we were too chicken to use the focus knob. On April 23, 2009 the telescope mirror was moved 40 microns (or 1.5 thousandths of an inch) and tilted 0.0072 degrees to arrive at the desired place. After two months of commissioning, science operations began on May 13, and the floodgates opened. Already, in the engineering data (now called *Quarter 0* data) there were lots of interesting observations of stellar brightness variations, and other interesting signals, to be seen—including signals from likely planets.

Science observations continued for a month, when in mid-June the spacecraft unexpectedly entered its *safe mode*. This was the first of two safe mode events that occurred in short succession. This state is triggered by the satellite's internal health monitors. If a parameter like temperature, acceleration, or its pointing relative to the Sun exceeds a predefined threshold, the camera turns off and the telescope reorients itself, both to maximize the power reaching the solar panels and to point its communication antenna toward the Earth. These safe modes take a day or two to diagnose, resolve, and then to get the telescope operating again [20].

The first safe mode occurred just before a planned *quarterly roll*, where we would normally have had a break in the data as the spacecraft rolled to reposition the solar panels and the onboard data was downlinked to Earth. We took advantage of the situation to execute these tasks early. The second safe mode, by contrast, occurred only two weeks later. Aside from adding large gaps in the time series of observations, these occurrences cause glitches (or *artifacts*) in the data as the telescope comes back online—the temperature changes across the spacecraft affect the sensitivity of the camera. Having two safe modes within a month was an inauspicious start. Fortunately they didn't persist at that rate as the onboard software was modified to make the telescope a bit less skittish. Ultimately, there were only eleven such events throughout the duration of the mission.

As data poured in through the summer of 2009, a portion of the science team began their feverish work sifting through the results. This work meant manually reviewing the threshold crossing events, sorting them based upon their promise as planet candidates, sending good candidates out to be observed, and following up those observations with additional telescope time, depending upon how the target fared along the way. Part of the rush to extract objects of interest from the data and to feed them into the observing program had to do with where the stars in the Kepler field are located on the sky. The Kepler field is visible primarily from March to October each year—with some portions that protrude into adjacent weeks of the calendar. The Kepler data were touching ground during the best time to make follow-up observations—when those stars were high in the sky throughout the night. If we couldn't get the necessary observations done before mid-fall, there would be no planet discoveries from the mission until the next spring when the Kepler field was again visible in the night sky.

Had the mission launched with the previous date in November 2008, there would have been a few months in the winter of 2009 to examine the data before the observing season began. With the March launch, however, the Kepler field was already visible, and the window was closing to make use of the time it could be seen. Obviously, the spacecraft shouldn't have launched before it was ready, but examining the ramifications of these kinds of delays gives interesting insight into the complexity of the scientific endeavor. Given how hectic things were following the spring 2009 launch, one can imagine the scramble that would have occurred with the third-to-last launch date of June 2008— where only a few weeks would have remained in the observing season as the data reached Earth.

It was clear from the few communications that made it through the grapevine to the science team that people were extremely busy. They were spending long days finding targets to observe, the nights observing them, and repeating the same thing the next day—in addition to analyzing the data they had just taken. It was also clear that the new members of the team, the participating scientists like me, had no idea what was going on. The newly expanded team still hadn't figured out how to communicate effectively across the collaboration. Information was meted out on a need-to-know basis. If you wanted something, you could request it, and the powers that be would get back to you.

In one enlightening email exchange this is exactly what happened. I requested access to the data. The response came back asking which data I wanted and why. In some circumstances, this would be a reasonable request, but when working with large astronomical surveys you often don't know what is interesting until you've had a chance to poke around a little. You need enough data to quickly test different ideas, discard some, and set aside others for future scrutiny. Maybe looking at the shape of a galaxy and the kinds of stars it contains is interesting. Or maybe it isn't. Maybe the transits of planets orbiting a particular type of star will give some useful insight. Or maybe they won't.

Kepler had observations of tens of thousands of target stars, and already a few hundred interesting planet-looking signals. There was no way to know which targets, or collections of targets, were going to be worth examining in more detail until we had a good idea of what the possibilities were. You can't prioritize one thing over another until you have a large enough sample to make comparisons. The need-to-know approach would have resulted in a large numbers of requests for information about individual targets, while missing some important findings because of an imposed form of tunnel vision.

I learned later that this had been the nature of communications through much of Kepler's developmental phase. Information didn't circulate among the team members unless it was necessary, and one needed permission from the top to share information with other members. This can make for slow going, especially when the people at the top are really busy. Playing devil's advocate, I can understand why one wouldn't want too much information to circulate when proposing a mission to NASA. The probability that your mission gets selected depends upon the unique strengths that it brings to the table. If too much information about your project circulates, it could be used by competing designs and diminish the chances of your mission being chosen to fly. Regardless of the incentives at play during development, as you transition from proposing, to building, to operating a mission, the approach to communication needs to adapt to the changing situation—as we were learning by experience.

For the participating scientists, our scientific projects were contracted with NASA directly, rather than with the Kepler mission itself, and our grant contracts lasted for two to three years at a time. This meant that, if we had only limited access to the Kepler data, our ability to fulfill our obligations was similarly limited. With Kepler data now on the ground, we all knew (or heard) that many science team members were exceedingly busy and had little opportunity to bring us rookies up to speed. But there was no good way for us to relieve any of the pressure, to assist with examining the data, or to provide information that would help effectively allocate telescope or personnel resources. All the while, there was the need for us to make good on our commitments to NASA.

This state of affairs persisted for a few months until the late summer of 2009. It so happened that Eric Ford, another participating scientist, and I were serving together on a NASA review panel in Washington DC. These panels convene every year to review and rate research proposals that are submitted to different NASA programs. Aside from the two of us, the panel also included two members of the broader Kepler team who oversaw important aspects of the mission—Edward (Ted) Dunham, the lead of the Kepler Science Team, and Patricia (Padi) Boyd, the Kepler mission's program scientist. The science team lead manages the body of scientists working on the mission, and the program scientist is the scientific liaison between a mission and NASA headquarters. Neither of these two people led the mission itself, but they were both high enough in the Kepler organization to possibly help with our dilemma.

At one lunch, Eric and I approached these two leaders to discuss the situation. The scientific collaboration that Eric and I viewed as a model for how these organizations should function was the highly successful Sloan Digital Sky Survey from the preceding decade.

That ground-based campaign to observe millions of galaxies and stars seemed to have successfully threaded the needle of giving people sufficient freedom of movement, open communication, and data access without resulting in chaos. When it came to Kepler, we felt stuck twiddling our thumbs with nothing to do, unable to help or to produce our promised results, while our colleagues at team meetings discussed their sleepless nights and blacked-out schedules with nothing on the horizon but work for the mission. The team lead was surprised by our situation, stating that he and the people he worked with were swamped and just trying to keep their heads above water. The idea that there were warm, but idle, bodies available to contribute was unexpected. Following our discussion, the two agreed to see what they could do to loosen things up.

The solution turned out to be straightforward. It was mostly a matter of sharing with the broader science team the phone numbers of the regular teleconferences for the follow-up observing program and the team of scientists who produced the planet candidate list by vetting the threshold crossing events. With those phone numbers, the participating scientists could join the conversation and start contributing their expertise and time to the tasks at hand. Throughout the late summer of 2009, access to the telephone lines and to the data continued to expand.

From my perspective, having been part of the science team for only one year, this change seemed like an important improvement, but not particularly groundbreaking. I didn't know any better. Nor did I fully appreciate the difference these changes made until, a year or two later, I was talking with a long-time senior member of the Kepler team. He said (in positive terms) that communication had always been a bit of a challenge, but then "the participating scientists went rogue" and changed everything.[2] Over the last few months of 2009, the entire character of the science team morphed into something much more collaborative. Various working groups coordinated their efforts and shared their needs, plans, and findings with each other.

---

2. This was only a few years after the 2008 US presidential election, where "going rogue" was all the rage.

The Multibody Working Group, with about half of the participating scientists, was especially productive in this new mode of operations.

As the first observing season drew to a close, an important deadline loomed ahead. Each January, the American Astronomical Society (or AAS—prounounced "double A S") has their largest meeting. Thousands of astronomers from around the world come to the United States to share their results. The January 2010 meeting was particularly important since it took place in Washington DC, near all the power centers of the US government, and where NASA is headquartered. Time was short for the science team to make exoplanet discoveries from Kepler that we could unveil at this upcoming meeting.

However, we couldn't just declare that our best candidates were planet discoveries. A significant concern for the mission was making false claims. In the early days of exoplanet science, planet retractions were not rare. By the time the 2010 AAS meeting took place, there were already sixty exoplanets (out of roughly four hundred) that were classified as "unconfirmed, controversial, or retracted." That is more than one in seven exoplanet claims that were dubious—an embarrassingly high number. The repeated mantra at early meetings of the science team was that the Kepler exoplanet catalog should not have any "holes." We should vet the planet discoveries so thoroughly that there would not be any planets removed from our list. A consequence of this approach was that the team was cautious in announcing new planets—especially as the mission began. If we were to claim an exoplanet discovery, it couldn't be a marginal detection—the evidence had to be overwhelming.

Given this intense desire to not have any holes in the Kepler catalog of planets, Kepler's first exoplanet discovery was Kepler-4b.

Now it may appear that starting with Kepler-4 instead of Kepler-1 was inconsistent with the stated wish, but the reason for this choice was mundane—there were already three known exoplanets in the Kepler field, TrES-2, HAT-P-7, and HAT-P-11. These three planets were assigned Kepler numbers 1, 2, and 3. Kepler-4 was the first previously unknown exoplanet in the Kepler field. It is the size of Neptune, but on a three-day orbit about its host star. At the time, Kepler-4 was among the smallest exoplanets yet discovered, since most of the early exoplanets were Jupiter sized. Indeed, in the first batch of Kepler exoplanet discoveries (Kepler-4 through Kepler-8), the remaining four planets were Jupiter sized and had short orbits like most other hot Jupiters [21, 22, 23, 24, 25].

Note that the traditional naming convention for exoplanets is that the star has the name, and the planet is designated by a letter following the name, beginning with the letter "b" and proceeding in the order of discovery, such as Kepler-4b. However, this is one of those instances where we scientists aren't good examples and don't practice what we preach. In casual conversations, we typically just ignore the star and refer to the planet as "Kepler-4," unless the situation necessitates more explicit terms. I will adopt this scandalously casual language in this book as well, unless necessity—or a nice turn of phrase—dictates otherwise.

The first batch of discoveries, four hot Jupiters and a hot Neptune, weren't revelations that really shook the field of exoplanets. It was a respectable catch given the constraints imposed by the launch date and observing season, but these planets were more of the same fare that astronomers had already sampled. What the early results did show, however, was how amazingly good the Kepler data were. The Kepler data were far beyond the quality of any existing instrument.

For example, one of the first targets studied by the science team was data from the previously known HAT-P-7 system. When the raw data from the commissioning phase of the telescope first appeared on the computer screens at NASA Ames, three scientists (Doug Caldwell, Jon Jenkins, and Natalie Batalha) saw the unmistakable dip in the brightness of the target, not as the planet transited the star, but as the light reflecting from the planet was blocked when it passed behind the star. That signal was as small as a transit from an Earth-sized planet—showing that Kepler's photometric precision would be good enough to see what it set out to see. The trio were so excited by what they saw, they immediately shared the news with the mission managers and with William Borucki. (William, in turn, knowing that the managers wouldn't miss an opportunity to brag to the public about the mission—and that the scientists would need to take time from what they were doing to prepare an analysis to go along with it—responded by saying, "don't tell the managers.")

The outcome was the first major result from the Kepler mission, a short paper in *Science* magazine that showcased the quality of Kepler's data, appearing only a few weeks after the initial observations touched the ground. Their analysis showed more than just the planetary eclipse. As light from the star reflected off the planetary surface, a portion of that light was reflected toward the Earth. As the planet orbited the star, the profile we saw from that reflected light changed. The planet went through phases exactly like the phases of the Moon. The Kepler data are so good that we readily saw the effects of the changing phases of the planet orbiting HAT-P-7.

But that's not all that the Kepler data could do. A second paper on the HAT-P-7 system appeared at the January 2010 AAS meeting, along with the announcement of the first five Kepler planets. In this work, a pair of astronomers from San Diego State University, William (Bill) Welsh and Jerome (Jerry) Orosz, teased out a more subtle effect from the data, called *ellipsoidal variations*. Here, the gravitational influence of the planet on the host star distorted the shape of the star like an ocean tide. Tides, in the astronomical sense, result from the difference in the strength of the gravitational attraction between the near side and the far side of one object due to the presence of a second object. The Moon raises a tide on the Earth because the near side of the Earth is attracted more strongly to the Moon than the far side of the Earth. The result is that the water on the Earth bulges slightly above the surface at two locations that point roughly toward and away from the Moon.

The same thing happens to a star hosting an orbiting planet. The star is slightly distorted into the shape of a rugby ball, or a *prolate spheroid*. That tidal distortion is always pointed in the direction of the orbiting planet. This means that, as the planet orbits, we observe the changing profile of the distorted star as it tracks the position of the planet. If the planet is along the line of sight, we are looking down the pointy end of the rugby-ball-shaped star. As the planet orbits, we eventually see the star from the side, then the point, then the side again. The changing profile means there is a difference in the size of the stellar cross section that we see. The changing amount of surface that we see produces a difference in how bright the star appears throughout the planetary orbit—these are ellipsoidal variations. The Kepler data are so good that you can see these ellipsoidal variations in the light curve— the effects of the distorted shape of the star caused by the gravitational influence of the orbiting planet [26, 27].

In fact, the Kepler data are good enough to see even smaller effects. When an object drifts toward or away from you, the wavelength of the light that it emits changes through the Doppler shift. We've seen this already in how the first exoplanets were discovered by looking at the Doppler shift of the lines in a stellar spectrum. A more subtle effect is that, because the wavelength of the emitted light is shifting, the energy that we measure for each photon shifts as well. When the star is approaching, the photon energy gets boosted slightly and the star actually appears brighter. When it recedes, the photon energy is slightly lessened and the star appears dimmer. This effect is large when the speed of the object approaches the speed of light, but the motion of the star caused by the gravitational pull of the planet is only a few meters per second, a tiny fraction of the speed of light. Nevertheless, the Kepler data are good enough to see this small, Doppler-boosting effect as well.

Another major result from Kepler appeared at that January AAS meeting—a result not directly related to exoplanets. This was the first result from the large group of scientists who studied asteroseismology with the Kepler data. By studying how a star shakes or pulsates, you can measure the properties of its interior. There are a few different ways that stars oscillate, and these oscillations produce changes in the star's brightness—sometimes a little, sometimes a lot. For example, as sound waves propagate through the Sun, they reflect off different layers in the Sun's interior (the core, the radiation zone, and the convection zone). The conditions in these layers, such as the density and pressure, make it so that certain oscillation frequencies will persist in the Sun's interior while other oscillation frequencies quickly dissipate. The Sun's brightness changes along with those persistent oscillations, and by studying them we learn about the properties of its layers and of the Sun as a whole—its size, mass, density, temperature, age, etc.

The ability to examine distant stars in this same way was showcased by a study of the star in the HAT-P-7 system, which already showed the secondary eclipse of the planet, the planetary phase curve, ellipsoidal variations, and Doppler beaming. With the Kepler data, these scientists measured the radius of the HAT-P-7 host star to an accuracy of one percent—ten times better than any previous measurement. Keep in mind that HAT-P-7 is six trillion miles away. At that distance, if you tried to measure the size of the star directly you would fail—it only subtends (or covers) an angle of four-millionths of 1 degree. We have no instruments with that capability. These astronomers were able to measure the stellar size to a precision of one percent with just over one month of data from Kepler. This first result was just the tip of the iceberg. We will return to more discoveries from asteroseismology later.

These early results were good reasons for astronomers to look forward to more data, and more announcements coming from the mission. The one-year anniversary of the Kepler launch was approaching. Kepler scientists were still combing through the first few months of Kepler data, checking for artifacts that would mimic a planet, eliminating spurious signals from eclipsing binary systems, and trying to find suitable new planet candidates to confirm as new planets. Winter 2010 turned into spring, and the Kepler field rose again in the night sky, beginning another season of ground-based observations.

Another international exoplanet conference was held in the Austrian Alps in late April 2010. There was a lot of anticipation for new results from Kepler going into this conference. With a sizable fraction of the exoplanet community gathered together, the talks from the Kepler mission . . . discussed the same five exoplanets they had announced four months earlier. There was nothing new. It was clearly a letdown for the participants, and for the discipline as a whole. With no new information presented at the conference, and no new scientific papers in the literature, frustration with the slow trickle of Kepler results got to the point that there was talk of boycotting talks from Kepler scientists until we produced something. It was bad, and would get worse.

The Kepler mission was slated to release its data to the public on a set schedule—roughly every year. The amount of data released each time was supposed to be half of the total data available. The reasoning here was that the mission was to supply a list of planet candidates, and we couldn't claim a planet candidate until we measured three transits. That, in turn, required data to be gathered for at least twice the orbital period of any planet detection. A planet with a one-month orbital period required at least two months of observations, catching a transit at the very beginning, middle, and very end of the time series. A planet with a two-month orbital period took at least four months of observations. Thus, using the originally planned schedule for releasing data, the planet candidate lists would be complete up to some maximum planetary orbital period allowed by the data.

One can argue about whether or not this approach was good, but that was the plan and those were the primary reasons. The first catalog, or *data release*, was scheduled for June 2010, six months after the AAS meeting in Washington DC and two months after the conference in Austria. With the way the launch date aligned with the observing season, and with some last-minute changes to the mission before launch, the development of some of our analysis software had been delayed. By the time the first data release was slated to occur, the science team hadn't completed the vetting process for many of the objects of interest. We weren't ready to release all of our results as the June 2010 deadline approached.

A major concern, given the history of retractions in the field of exoplanets, was that releasing all of the data without sufficient examination could shed bad light on the mission as a whole, and further mar the perception of the discipline. Not only would people draw conclusions that weren't justified, but there may be some who would comb through the data and make claims for Earth-like planet discoveries that were simply false positive signals that hadn't yet been addressed with our follow-up observations. I, personally, think that these concerns were well justified given the frequency of retracted exoplanet detections up to that point and the tendency for the news media to sensationalize stories.

It's not hard to imagine a story blowing up in our faces. Kepler was the leading scientific endeavor in a discipline with a lot of public interest. An announcement by a third party of the discovery of a potentially habitable world using "Kepler data" would be a major headline. A subsequent retraction would make a similar headline, and the fact that the false claim was by an unaffiliated group would not likely be part of the headline. So NASA, Kepler, and the scientists on the mission could easily be the ones to suffer the loss of credibility in the eyes of the public.

Given these concerns, we petitioned NASA to withhold data on 400 of the 150,000 target stars from the release of data that spanned Quarter 0 (the engineering data) and Quarter 1—each roughly one month in duration. The choice to "sequester" these targets didn't come lightly, and it required some introspection on the part of the science team, as well as some explaining to NASA about why the Kepler mission should be able to deviate from its agreement. I helped draft the document that circulated among the powers that be in Kepler, and that ultimately made this case to NASA. At the time there were just over 700 identified planet candidates. Our proposal kept back data on just over half of them until the next scheduled data release—eight months hence [28].

Looking back years later, I think it fair to say that our worst fears did not materialize, and likely wouldn't have. After we had released a few of these catalogs to the public, and after a lot of high-quality results came from the broader community, we voted to change our policy about releasing data. We dropped the "release half the data with each new catalog" idea and published it almost as quickly as we could bring it in and process it. This later approach of releasing data as soon as possible was adopted by Kepler's most notable successor mission, TESS (the Transiting Exoplanet Survey Satellite), which we will discuss at the end of this book. This policy might not be viable in some, or even most, cases—giving scientists time to look at their observations can have its benefits. But, for the later part of the Kepler mission and the TESS mission, the policy seems to be a net positive.

Nevertheless, given our inability to see the future, and the plausible concerns for the reputation of the mission, our proposal to sequester data on 400 targets was accepted by NASA. News of the sequestered data received mixed reviews. Among exoplanet scientists there was some justified disappointment with the delay. I think, for the most part, that people understood the issue. Regardless, the data that were released were sufficient to keep a lot of people busy with interesting projects for a long time. Even now there are interesting systems identified in the first catalog that have not been fully explored.

Included in the data we released to the public in June 2010 were the light curves of virtually all of the 150,000 target stars, and over 300 planet candidates—several of which were in interesting systems for further study. Among them, for example, were five systems that showed multiple planet candidates orbiting their host stars. Such multitransiting systems would play a central role in Kepler's scientific output in the coming years, and I was fortunate to lead the paper announcing these first-of-their-kind planetary systems.

For the news media and some conspiratorial parts of the general public, the response to the data sequestration was less charitable. NASA had a mission with the capability of detecting planets that could host alien life, and NASA suddenly decided to withhold a bunch of those data. A number of opinion pieces, blog posts, and news articles ran on this issue—often with headlines that were more sensational than the text. There was never any question about whether or not the sequestered data would see the light of day—all NASA data eventually get released to the public. The only real question was when it would happen. That wasn't good enough in the eyes of a few.

One incident caused a lot of wasted time for many members of the Kepler Science Office at NASA Ames. Dimitar Sasselov, a science team member from Harvard, gave an innocuous TED talk shortly after the data were released. He estimated from those early data that Kepler would eventually find somewhere around sixty Earth-like planets. Well, that blew up. People submitted Freedom of Information Act (FOIA) requests to see all communications among the science team about these planets and the decision to sequester some of the data. The director of NASA Ames had to craft a statement basically saying that the opinions expressed in the talk didn't reflect the opinions of NASA. Science team members were told in an email to watch what we say because "there is

intense public interest in Kepler and little news to grab onto, so people are reading the tea leaves and paying more attention to details than
we might imagine." And the scientists who were already overworked
trying to get the Kepler data vetted and out to the public spent several
days combing through their emails to comply with the FOIA request—
providing essentially the same information that you, the reader, saw a
few paragraphs ago.

While the sequestration of the Kepler data made a bright flash, it
was (fortunately) short lived. It didn't take long before there was more
important news from Kepler to fill discussions in the hallways and
surrounding the water coolers. A year later, when data on all of the
targets were available, and new results were pouring in as quickly as
the journals could publish them, most of the concerns about holding
back some of the early data were forgotten. In one instance, I was eat-
ing lunch at the first Kepler Science Conference, held in December
2011 at the NASA Ames campus. I was chatting with someone at my
table who turned out to be a science journalist. We were talking about
all of the great results from Kepler that were being presented at this
conference.

The subject of the sequestered data came up and he asked my opinion. I told him that I understood the science team's perspective and
helped draft the initial document on their behalf to keep those data
proprietary for a bit longer. I also said that I felt the hand-wringing
and the media outcry was overblown—something along the lines of
"A lot of scientific data never get released to the public, and this was
a six-month delay. Many of the people who were complaining about
it haven't done anything with the data even now. And all these media
stories were about how the Kepler team was doing something nefari-
ous, and how NASA was keeping secrets. It was six months." (It was
actually eight months, but I only worked out the dates as I write this.)
His response was "Yeah, I think I wrote one of those stories." Then, we
continued our conversation about the new results from the mission.

4

Hot Jupiters and Hot Earths

Immediately prior to the June 2010 release of the redacted Kepler catalog, the science team met in Ebeltoft, Denmark, a small town on the coast of the Jutland peninsula and a few miles from the University of Aarhus, where the Kepler Asteroseismology Science Consortium was headquartered. For years, our European collaborators regularly traveled to the United States to attend team meetings in California or Boston. This was an opportunity for them to host us. In my opinion, this meeting in Denmark was the most pivotal meeting of the entire mission in terms of its scientific impact. Not only was it a prelude to our first planet candidate catalog, but a host of other scientific investigations were either presented in preliminary form to the science team, or initiated by different members as a consequence of the meeting.

It was a hectic time for everyone getting ready for this meeting. Most of the team were neck deep preparing the catalog for publication. This was also the time when I was preparing the paper that announced the discovery of the five multitransiting systems, which was a significant milestone for the field of exoplanets and which was to coincide with the first planet candidate catalog. I had spent the prior week in Mexico attending a scientific conference about dark matter. This was an area of research that I picked up while I was at Fermilab, a national lab near Chicago. (Incidentally, through this research I had the opportunity to work with the world-renowned cosmologist that eluded me when I was a graduate student. Craig Hogan had moved from Seattle to Chicago to become the director of the Fermilab Center for Particle Astrophysics, where I was already working.) I took advantage of one day at the dark matter conference, where I needed to stay close to the bathroom, to get the "five-multis" paper far enough along to send it up the Kepler chain of command for approval. In the meantime, others on the science team were preparing the planet candidate catalog that was about to land.

By the time we arrived at the meeting everyone was exhausted. Nevertheless, there was still work to be done on both papers, and time was short. There was one person on the team who was the lynchpin of the whole enterprise, Jason Rowe (now at Bishop's University in Quebec, Canada). He had been worked to the bone in the weeks and months leading up to the meeting, as his analysis of the spacecraft data was crucial for both the catalog and the five-multis papers. Upon setting foot in Denmark, he vanished. No one had seen him since leaving the airport. No one knew where he was, he didn't answer any emails, and knocking on his door produced no response.

We worried that something bad may have happened to him. The situation conjured up memories from a few years prior. Jason's graduate advisor, and leader of an important Canadian space mission, had attended a scientific conference in Washington DC. He too mysteriously vanished because a late-night altercation had put him into a coma for several months. Now, in this foreign country, with no sign of his former student, there was definitely some concern among the team members. We exhaled a collective sigh of relief when we learned that he had simply been asleep—for thirty-six straight hours.

During the Denmark meeting, we discussed a variety of interesting Kepler systems—ones that became icons in the field: Kepler-9, Kepler-10, Kepler-11, KOI-54, and KOI-126, to name a few. We'll discuss these systems in more detail shortly; just note that for over a decade, almost every scientific talk, poster, public lecture, advertisement, and classroom slide used artists' renditions of these systems—including from scientists unaffiliated with the mission. Ultimately, these systems became the seeds from which grew many subsequent, large-scale investigations.

When these systems were introduced to the science team, it was by their designation as Kepler Objects of Interest, KOIs, rather than their

Kepler planet designation. This was and remains a source of some confusion (at least to me) after the results were announced to the public. All
of our analysis, the follow-up observations, and paper authoring were
accomplished using only the system's KOI number as a reference. The
planet designation wasn't assigned until after the peer review process
was complete and the paper was accepted for publication. That process could take months, and usually only involved the lead author and
a few coauthors. By the time the planet numbers were assigned, most
of us would have moved on to something else. For myself, and many
others, it would be years before we started thinking of these systems as
Kepler-9 instead of KOI-377, or Kepler-11 instead of KOI-157. Even
now, after more than ten years, it is only the really popular systems
where I've learned their Kepler designation instead of their KOI designation. I often look up my own planet discoveries to remind myself
of what Kepler planet number they were assigned.

About halfway through the Denmark meeting we had a discussion
about KOI-72. This system caught our attention because the star was
similar to the Sun, it was bright (among the brightest in the Kepler
survey), and the innermost planet was small—only about fifty percent
larger than the Earth. While there remained some work to nail down
its properties, KOI-72 could be the first rocky exoplanet discovered
orbiting a Sun-like star. That was a big deal, since planets that similar
in size to the Earth had not been seen before. Our discussion turned
to who would lead the effort—one of those uncomfortable situations
where people have their opinions, but don't want to speak up for fear of
hurting others' feelings. Collectively, we wanted everyone to have the
opportunity to lead a study on something exciting, but also wanted to
reward those who had devoted considerable work over many years into
making the mission a success.

I don't have much affinity for awkward times like these, so I blurted
out what I thought most everyone else was thinking, or at least something few would object to: "I nominate Natalie." And so it was that
Natalie Batalha led the effort to confirm the discovery of what became
the Kepler-10 planetary system. (Of course, I'm probably giving myself
too much credit as it is quite likely this decision had already been made

and that the discussion was a formality. But the brashness of my relative youth and inexperience makes me want to cast all my actions as being part of the natural growth of an intrepid young scientist.)

Kepler-10b was the first rocky exoplanet confirmed by Kepler, and the first "definitively" rocky exoplanet seen orbiting a Sun-like star. I'll note a bit of hair-splitting here. Another exoplanet, CoRoT-7, has similar properties and is probably rocky [29]. It was discovered in 2009, but its mass was sufficiently uncertain that it left open the possibility of it having a substantial atmosphere. This meant that the overall density of CoRoT-7 could be as little as half that of Earth—that is, not rocky. The density of Kepler-10, on the other hand, is at least twenty percent more than the Earth, so it must be rocky. If you read carefully the press releases and statements by Kepler scientists at the time, you will notice the delicate language given by the members of the science team, because of this competing claim.

A notable characteristic of Kepler-10, beyond its rockiness, is the planet's orbital period. Kepler-10 orbits its host star every twenty hours. Twenty hours! Its year is less than one Earth day. Its orbit is so close to the host star (at a distance of only three and a half times the stellar radius) that when it is high noon on the surface, with the star directly overhead, the star would appear three thousand times larger than the Sun appears in our sky. It is so close to its host star that the surface of this planet on the side directly facing the star would be a sea of molten rock. Yet, because a planet that size and that close to a star is unlikely to retain an atmosphere, it cannot transfer that heat from the sunlit to the shadowed side of the planet. Thus, the temperature of its dark side is likely well below freezing. Plate 5 shows an artist's rendition of the planet Kepler-10b.

Beginning with their unexpected discoveries in the mid-1990s, hot planets like the hot Jupiters and Kepler-10 had been constantly turning up in Doppler and transit surveys. Indeed, hot planets will dominate most early discoveries with these surveys because they have both the largest and the most frequent signals. However, to get a good idea of how common these systems really are, we need to account for the fact that they are easy to find. When we make this accounting for the case of hot Jupiters, we find that they orbit roughly one percent of stars. Short-period rocky planets like Kepler-10 turn out to be similarly common, orbiting one to a few percent of stars. These rocky planets, however, don't appear to form in the same way as the Jupiters.

When hot Jupiters were first discovered, astronomers proposed a number of theories to explain their existence. It was clear they must have a different history from the solar system. In the solar system, the gas-giant planets Jupiter and Saturn formed quickly from the large, gaseous disk of material that surrounded the Sun. Only at large distances from the nascent Sun were the temperatures of this proto-planetary disk cool enough for a wide range of compounds to condense from that disk. Closer to the Sun, the temperatures in the disk were higher, and more of the material remained in the gaseous state, unable to condense to form planets. At the same time, if you got too far away from the Sun, the disk had less mass, and the timescales for planets to form were much longer.

The number of orbits that a planet makes is a useful clock to measure planet formation. More distant planets orbit more slowly, and take longer to sweep through the material in the vicinity of their orbits. The Earth orbits once per year, Uranus's and Neptune's orbits are closer to a century, so they have orbited the Sun a factor of a hundred fewer times than the Earth. The disk material is also more diffuse that far away from the Sun, meaning there is less stuff available to build the forming planets. Jupiter and Saturn, by contrast, have relatively short orbits of a decade or two and a lot of material from which to form. Once their solid cores gained enough mass, they started pulling gaseous material directly from the protoplanetary disk and onto their surfaces to make their substantial atmospheres.

How Jupiter and Saturn developed their core is the subject of ongoing debate—whether through the gravitational collapse of a dense region of the disk, or through the collisions of small asteroid-sized objects, or (in a more recent theory) through the accumulation of streams of tiny, gravel-like pebbles that flow within the disk. Either way, once their cores were established and they started accreting disk material, the larger they became the more quickly they could grow. In the end, the gas giants are massive (Jupiter is three hundred times the mass of the Earth, and Saturn a hundred times), and that mass is dominated by the hydrogen and helium that they pulled from the disk. Since the gas giants gain most of their mass while the disk is present, they must form within the few-million-year lifetime of the disk.

Meanwhile, in the outer solar system beyond Jupiter and Saturn, light compounds like water, methane, and ammonia start to condense out of the disk and accumulate in the atmospheres of the planet cores in that region. These light compounds are made with some of the most abundant elements in the universe: carbon, nitrogen, oxygen, and hydrogen. When they condense, they too can form planets with substantial masses—like Uranus and Neptune. These *ice giants* are smaller than their gas-giant counterparts, being about one-tenth their mass and half their radius. Comparing these planets to Earth, the gas giants are hundreds of times the mass of the Earth and ten times the radius, while the ice giants are fifteen to twenty times the mass and about four times the radius.

The situation is different in the inner solar system, however. Here, there were numerous, small rocky objects that never grew to sufficient size to start the runaway growth to becoming a gas giant. Throughout the time when the disk is present, they are embedded within it, and the disk drives them onto mundane, circular orbits. (If their orbits are eccentric, meaning that they are elliptical rather than circular, they plow through the gas as they approach and recede from the Sun, friction dissipates that eccentricity from their orbit, and their orbits circularize.) Once the gaseous disk disperses, however, it no longer dampens the eccentricity of their orbits, and their eccentricities start to grow. Eventually the orbits of these different planetesimals cross, causing collisions and mergers among the bodies. Over a few hundred million years, the rocky planets grow to the sizes they are now. It takes about a hundred times longer to form terrestrial planets (hundreds of millions of years) than it does to form gas giants (millions of years).

For decades, the scientific consensus was that gas giants form in the outer solar system, beyond the *ice* or *snow* line where water and other light compounds can condense. So discovering Jupiter-like planets with orbits of only a few days required some explaining on the part of planet-formation theorists. How do you get a planet like Jupiter to form in a place where most of the materials that it is made from cannot condense to form a core around which to accrete an atmosphere? With this challenge set before the theorists, a few broad hypotheses soon surfaced.

One idea is called the *gravitational instability model*. Here, a region of the protoplanetary disk with a density slightly above the average will gravitationally attract neighboring material. As more material coalesces, the region becomes even more over dense, which attracts still more material. This positive feedback loop quickly generates clumps of gas that are sufficiently dense to collapse into planets under their own weight. Computer simulations of this model often produce a handful of planets on short orbits. Planets from these systems would interact strongly with each other, sometimes smashing together, sometimes exciting each other's orbits so that they crash into the central star or are completely ejected from the system. A drawback of this model is that it requires a very massive disk to get the collapse to occur in the first place, and most observed disks don't have the required mass. So, while this process can occur, and likely does occur in some cases, it does not appear to be the primary means to form hot Jupiters.

A second hypothesis for hot-Jupiter formation posits that the planets form the same way and in the same location as Jupiter and Saturn, but then migrate toward the central star. This core-accretion plus migration model explains the existence of hot Jupiters by first ignoring the problem and then kicking it down the road to deal with later. With the planets forming the same way as Jupiter and Saturn, we need only explain how they then moved from the outer parts of the planetary system to deep inside the inner part—before somehow coming to a halt at one-hundredth of their original orbital distance and just prior to hitting the central star. Here, again, a few options appeared to explain the migration—disk migration and eccentric migration.

In disk migration, the planet's interaction with the surrounding disk causes it to lose energy and drift toward the star. Its migration eventually stops when the disk dissipates, leaving the planet at its observed orbital distance along with any other debris that it entrained on its journey. Eccentric migration, on the other hand, occurs when a planet is excited into a highly elliptical orbit. Here, when the planet plunges toward the central star, the tides that the star raises on the planet cause the planet's shape to distort into an ellipsoidal, rugby-ball shape. (We saw earlier that tides in the HAT-P-7 system caused a similar distortion, but of the star.) The friction induced by the constant distortion of the planet converts energy from the planetary orbit into heat in the planetary interior. Ultimately, the planet's orbit circularizes near its point of closest approach—at the location where we observe them today.

With eccentric migration, the can has now been kicked down the road again. We still need a mechanism to excite the orbits of the planet in the first place. Theorists, being the clever devils that they are, produced yet another collection of ideas to explain the excited orbits. Two prominent ones are that the orbits of the planets could be perturbed into high-eccentricity states by a distant star, or that there might be strong gravitational interactions with other planets in the system that cause the eccentricity to grow.

Each fork on the road for explaining the origins of hot Jupiters— gravitational instability or core accretion, eccentric or disk migration, stellar or planetary perturbations—implies a set of correlated observations that we can make to test their validity, or rather the frequency with which each mechanism occurs. For example, the disk-migration scenario predicts the presence of additional planets in the system, whose orbits were tied to the gas giant during its migration. As the giant planet moves inward, small bodies embedded in the disk will be captured into orbits whose orbital periods form integer ratios with the giant. That is, the giant planet's slow migration will shepherd small objects into specific types of orbits, where the small planets orbit twice for each orbital period of the giant, or three times for every two—forming numerical ratios of 2:1, 3:2, 4:3, and so forth. These orbital configurations (called *mean-motion resonances*) are a natural consequence of the dynamical interactions that occur when orbiting objects drift inward or outward in a system.

We see mean-motion resonances throughout the solar system. They appear in the orbits of the moon systems of Jupiter and Saturn. They show up in the relationships between the orbits of Jupiter and a number of asteroids. And, most importantly for our purposes, it is seen in the 3:2 orbital pairing of Neptune and the non-planet Pluto. This last example is strong evidence suggesting that Neptune initially formed closer to the Sun (at roughly half of its present location), and later migrated outward in the solar system. When it moved outward, it plowed into a ring of comet-like debris called the Kuiper belt. Pluto, being the big fish in that pond, was captured into the 3:2 mean-motion resonance with Neptune that we see today. Pluto is actually one member of a family of *plutinos* that are trapped in similar resonances because of this past migration of Neptune.

If we were to find small planets near mean-motion resonance with a hot Jupiter, it would be a clear indication that the Jupiter was moved to its present spot through disk migration. Early searches for these small, resonant companions to a few transiting hot Jupiters turned up empty handed. A larger survey using Doppler measurements was also fruitless. With Kepler data, we had a uniformly sampled set of hot-Jupiter systems where we could make more definitive statements about the presence of these small, trapped planets, and as a consequence, about the importance of disk migration for hot-Jupiter formation.

I led a study of hot-Jupiter systems in the Kepler data trying to find these companions. The conclusion from these investigations was that hot Jupiters are lonely. In fact, the submitted title of my Kepler study was exactly that: "Hot Jupiters Are Lonely." But, to my chagrin, and the chagrin of my coauthors and colleagues, the journal said that I needed a title that sounded more professional. So, while the findings were the same—hot Jupiters are indeed lonely—the title was changed to something along the lines of "Kepler Constraints on Boring Word-Salad Text." We use the original title often in our talks and lectures, but that is just a consolation prize [30].

The lack of close companions to hot Jupiters disfavored the disk-migration scenario, but it didn't rule it out entirely. It was possible, even likely, that at least some hot Jupiters formed through this mechanism—the real issue was to know the fraction that did so. The first counter-example to the lonely hot-Jupiter paradigm was identified in 2015 when the Kepler telescope found two small planets, one on either side of the hot Jupiter WASP-47. Finally, after studying more than a hundred hot-Jupiter systems, astronomers found one with these tell-tale neighbors. While definitive proof is not really what science is about, this observation was strong evidence that disk migration does indeed play a role in the formation of some hot Jupiters. Since hot-Jupiter planets are found orbiting roughly one percent of stars, systems like WASP-47 correspond to only one percent of the one percent. I personally believe that the discovery of nearby companions to WASP-47 is one of the most important exoplanet discoveries to date. (My opinion may have to do with the fact that it overlaps with a lot of my personal research, but we can pretend that I'm not biased.) Some recent work suggests that the number of hot Jupiters with nearby companions may be a few percent.

For the remaining ninety-five percent or ninety-nine percent or so of hot-Jupiter systems that form through eccentric migration, there are observations we can make to distinguish the underlying causes that produced the high eccentricity in the first place. One common mechanism involves a distant star or large planet where the orbit of the distant object is misaligned with the orbit of the future hot Jupiter. Over time, the influence of the distant object will start to align the orbit of the Jupiter with its own. While this happens, the Jupiter exchanges the inclination of its orbit (with respect to the distant perturber) for a high eccentricity. Later, the cycle reverses, the orbit begins to circularize again, and its inclination grows. The orbit of the Jupiter oscillates back and forth between a misaligned, circular orbit and an aligned, eccentric one. (The underlying physical principle that produces the oscillation between these two states is conservation of angular momentum. Eccentric orbits have less angular momentum than circular ones. Relative to the orbit of the distant perturber, the misaligned circular configuration will have the same angular momentum as the aligned eccentric one.)

These cycles in the orientation and shape of the orbits of the Jupiters are called Kozai cycles, or Kozai–Lidov, or Lidov–Kozai, or von Zeipel–Lidov–Kozai cycles (or any combination thereof, depending upon how you want to distribute credit for their discovery). The final step in converting a regular Jupiter undergoing Kozai cycles into a hot Jupiter is to shrink and circularize its orbit. This is done through tidal dissipation—the friction in the planet that occurs when its shape is distorted by the tidal force from the star. That effect is strongest when the orbit of the Jupiter is highly eccentric. The result is that the planet's orbit will shrink down to a size that is comparable to the distance of its closest approach to the star. (If you work through the math, the planet will land somewhere between one and two times the close-approach distance, depending upon its initial eccentricity.)

Kozai cycles are not the only way to produce high-eccentricity orbits. Another common mechanism to excite the orbit of a Jupiter requires that multiple giant planets form within the disk. (Our solar system produced two, so it isn't hard to imagine a different system producing three or more.) Here, as the planets interact with each other gravitationally, the system can become unstable, and one or more of the planets can be ejected entirely. Ejecting a planet from the system excites the eccentricities of the orbits of the remaining planets to large values. From there, the story is familiar, with the high-eccentricity orbits shrinking and circularizing through tidal dissipation. There are other variations on these themes for how to perturb the orbit of the planet, but they all produce high eccentricities, and eventually hot Jupiters.

The different ways to make a high-eccentricity orbit bring different observable consequences. Consider the case where the hot Jupiter originates from a system of giant planets where one of the planets is ejected. In order to form a system with multiple gas giants in the first place, it would help to have additional heavy material from which they can form. We see from observations of stars with orbiting hot Jupiters that they have more metals in their atmospheres, implying more metals in the protoplanetary disk to build the planets. The observation that hot Jupiters are seen orbiting metal-rich stars is consistent with the necessary preconditions for forming multiple large planets. (Note that when an astronomer refers to a *metal*, they usually refer to anything heavier than helium. It's only when you press the issue that they will grudgingly concede that not all metals are metals. Some metals are nonmetals, and some metals are noble gases. For now, however, we'll just adopt the "anything heavier than helium" definition.)

In addition, for these systems, there may be distant planets that remain in the system—ones not involved in the ejection process. We see this too; many hot-Jupiter systems have additional giant planets on distant orbits. Finally, some of these systems might leave behind planets with highly eccentric orbits that have not yet circularized. (For eccentric orbits, the more distant the point of closest approach, the longer it takes for the orbit to circularize.) We see systems like these too, where some planets have orbits that are extremely elliptical, with eccentricities above ninety-five percent. In these cases, the circularization process isn't complete.

On the flip side of the planet-ejection scenario, we would expect the ejected planets to be out there somewhere. These free-floating, "rogue" planets should be roaming through the galaxy, untethered to any star. It turns out that we do see such planets. We can find them through the gravitational microlensing signatures that they induce on distant stars, and we occasionally see them because of the glow from the residual heat they have from their formation. These latter planets would be easiest to find in regions that are actively forming planetary systems—which is where we observe them. Some estimates derived from these observations place the number of these starless, rogue planets as comparable to the number of stars in the galaxy—hundreds of billions of them. Not all rogue planets are necessarily detritus from hot-Jupiter formation—some may simply form on their own—but their presence keeps the planet-ejection mechanism viable.

One interesting side note to consider for giant planets that are kicked out of their planetary system is the possibility that they might harbor life. In the solar system, we expect Jupiter's moon Europa to have an ocean of liquid water under its surface. That water leads scientists to consider Europa to be one of the more likely places where extraterrestrial life might exist. The heat that keeps Europa warm comes, not from the Sun, but from Europa's gravitational interaction with Jupiter and its other moons—tidal heating. The distortions of the moon interior from tidal flexing can generate a lot of energy—Jupiter's moon Io experiences the same heating, which fuels active volcanos all across its surface. Not only is the energy substantial, but tidal heating can persist for billions of years. A pair of interesting studies, including one led by one of my undergraduate students, Ian Rabago, looked at whether or not these moons would survive the ejection process, thereby preserving the energy source that prevents the liquid water from freezing.

The answer came back as yes. If Jupiter and its moons had been in a system where they were ejected by strong gravitational interactions with neighboring planets, there is a high probability that Jupiter would keep its major moons. And there is a high probability that the orbital relationship that generates Europa's internal heat would persist after the ejection. This gives the possibility that there could be life on a moon that is orbiting a giant planet that was ejected from its planetary system and is drifting through the galaxy with no host star—and that life could persist on that moon for billions of years [31, 32].

Let's return to the different ways that hot Jupiters could form. For those that might form through Kozai cycles, where the orbital inclination can vary over a large range, we might expect that their final orbits would be misaligned relative to the rotation axis of their host star. In extreme cases, Kozai cycles can be severe enough to cause the planetary orbit to become highly inclined—even flipping the orbit completely over. By contrast, in the solar system, where we haven't experienced these effects, the Sun's rotation axis is nearly aligned with the orbits of the planets. The Earth's orbit is misaligned with the Sun's spin by only about 7 degrees, as opposed to a nearly 180-degree misalignment that can occur from Kozai cycles. Measuring large misalignments between a planetary orbit and the stellar spin would support this mechanism for forming hot Jupiters.

It turns out that we can measure the relative angle between the stellar spin and the planetary orbits. One way is through an interesting signature called the Rossiter–McLaughlin effect. For this effect, you need to have a spinning star, and a transiting planet. When the star spins, you will have one side of the star that is approaching you and the opposite side that is receding from you (as long as you are not looking straight onto the pole of a star). During the time when the planet is not in transit, the spectrum of the star remains centered at its nominal value. However, when the planet transits across the star, it blocks a portion of the starlight. When it blocks light from the side of the star that is approaching you, you will see an anomalous Doppler shift. Because light from the approaching limb of the star is blocked while light from the receding side of the star is unobstructed, it will appear as though the star suddenly recedes from us. Later in the transit, when the planet passes to the receding side of the star and blocks some of that light, while the approaching side is unimpeded, the star will suddenly seem to be approaching.

If a planet is orbiting in the same direction that the star spins (as is the case with all of the planets in the solar system), it will block the approaching side of the star first, then the receding side. If the planet orbits opposite to the stellar spin, then the obstructions occur in the opposite order. And, if the planet orbits over the pole of the star, it will block only the approaching or receding side, or you won't see an effect at all. Observations of this effect show that a fairly large fraction (like thirty percent) of hot Jupiters have orbits that are highly misaligned, or even backward, relative to the spin of their stars. This observation is consistent with an eccentric-migration origin for many hot Jupiters—their orbits were highly excited in eccentricity and inclination and then circularized. (I'll briefly note that when the first five planet discoveries were announced by the Kepler team in January 2010, the spring launch and short observing season meant that ground-based observations were limited. Kepler-8 was teetering on the edge of *unconfirmed*, until a measurement of this Rossiter–McLaughlin effect bolstered its standing as a real planet. Those observations were taken near twilight a few days before the observing season ended for 2009.)

The story about the origins of hot Jupiters does depend on how you define a hot Jupiter. I usually think of them as gas giants with orbital periods less than one week, typically three days. The main reason for my affinity for this definition is that there appears to be a slight dip in the frequency of gas-giant planets near orbital periods of one week. Some theories for the formation of systems like WASP-47—the hot Jupiter with small nearby companions—also produce systems with similar properties, but where the gas giant remains on a slightly larger orbit. It may be that eccentric migration, from planet scattering or Kozai cycles, produces one population of hot Jupiters, while disk migration yields Jupiters with a wide range of orbital periods, including some extreme cases with orbital periods of only a few days.

If this scenario were true, where the short-period tail of the distribution of disk-migrating Jupiter-like planets spans a wider range than the few days that we see from eccentric migration, then we should see a population of warm Jupiters with nearby planetary companions and orbits of a few weeks. Systems like that actually made an early debut in the Kepler data. The paper that announced the first five multitransiting systems (the paper that accompanied the first release of the Kepler data) included an example of a system like this, which may have formed from disk migration in a fashion similar to WASP-47.

KOI-191 (Kepler-487) has a Jupiter on a fifteen-day orbit with three smaller companions on a sixteen-hour orbit, a two-and-a-half-day orbit, and a forty-day orbit. While the planet with the sixteen-hour orbit wasn't seen in the original data, this system was selected for that original paper because of the strange juxtaposition of the large planet sandwiched between two smaller ones. Systems like this and like WASP-47 can't have had a violent origin through eccentric migration, or the small planets in the systems would have been lost. They most likely arose from a quiescent migration within the planet-forming disk.

At the time, I didn't consider (and still don't) a Jupiter with a fifteen-day orbital period to be a hot Jupiter. This issue of definitions, that hot Jupiters have orbital periods near three days, is the reason I believe WASP-47 to be so groundbreaking. WASP-47, with its three-day Jupiter and three companions, was the counterexample that proved not all hot Jupiters have violent origins. Jupiters with orbits of a couple of weeks, whether they have small companions or not, may or may not have formed this same way. Some other mechanism might be able to explain them. However, WASP-47 being as extreme as it is, extends the limits of disk migration past few-week orbits for the Jupiters into few-day orbits—supporting a disk-migration history for less extreme systems with similar architectures. It is systems like WASP-47, exceptions to the rule, that give the best clues as to the relative contribution of these two formation channels for making short-period gas giants.

Regardless of the origin stories of any individual hot Jupiter, all of these observations clearly suggest that their histories are very different from that of the solar system. Most of them appear to have had troubled childhoods which drove them to their current locations, clearing out any planets in the inner planetary system that may have had the gumption to form. Very few hot Jupiters appear to have formed in place, or to have migrated to their current locations through interactions with a gas disk. From what we've seen, it appears that the bulk of the population of hot Jupiters form through some sort of high-eccentricity-plus-circularization mechanism. A separate formation channel makes warm Jupiters with small companions—with the short-period tail of these latter systems stretching into the hot-Jupiter domain, and producing systems like WASP-47.

Hot Jupiters aren't the only planets with short orbital periods—there is a substantial second population of planets with quite different properties. These are Earth-sized planets on roughly one-day orbits. Kepler-10b, with its twenty-hour orbital period, is a member of this second group. We had seen these planets prior to Kepler, but it took some time to appreciate what these hot-Earth planets were, and how they differed from more typical planets. The first hot Earth to be seen was 55 Cancri e. It was found using Doppler observations four years prior to the Kepler launch. However, it wasn't initially clear what its true orbital period was. With Doppler measurements, you can only make observations at night, and because this planet's orbital period is less than a day, measuring its orbital period was like trying to measure the speed of a spinning object that is illuminated with a strobe light. If the object is cycling faster than the strobe, you don't see the real rate of rotation. Instead, you'll see an alias of the rotation period caused by the interplay between the rotation period of the object and the time between flashes of the strobe. In the case of 55 Cancri e, it looked like the orbit was a couple of days rather than the true orbital period of about eighteen hours. It took several years to sort the issue out, which happened around the same time as the Kepler-10 discovery [33, 34, 35, 36].

Eventually, Kepler uncovered lots of hot Earths. However, these discoveries were hidden from us for two reasons. First, they were swamped in numbers by the other planets that materialized from the data. Second, we weren't really looking for them, and for those we found, relatively little was done by the science team to study them further. After all, why study Earth-sized planets on one-day orbits when there are similar planets on larger orbits, especially those within the habitable zone of some star? This situation changed when a group led by Saul Rappaport, at the Massachusetts Institute of Technology (MIT), started combing through the publicly released Kepler data on their own.

This team was looking for planets with really short orbital periods—only a few hours, and certainly less than a day. These were orbital periods that the Kepler data analysis pipeline didn't examine. With their search, they uncovered a remarkable system, what is now called Kepler-1520, but was then called KIC 12557548 (or KIC 1255 for short). This planet has an orbital period of only sixteen hours—incredibly close to the host star. The intense light from the star heats the planet to the point that it vaporizes the rocky material from the surface. Solid material evaporating from an object is normally what we see with comets. These small objects are made of volatile ices, and their small surface gravity makes it easy to evaporate the material they are made from, which then forms their iconic tails of ionized gas and dust. For KIC 1255, we have a whole planet made of terrestrial material that is nearly the size and mass as the Earth—with a tail of vaporized rock trailing behind [37, 38]. Plate 6 shows an artist's rendition of the planet in KIC 1255/Kepler-1520.

As this MIT group continued to find other small planets with short orbital periods, we, on the Kepler mission, got the message and adjusted our planet-searching algorithm to start looking for shorter orbital periods with our analysis software. The ultra-short-period planets (or USPs) that turned up are significantly smaller than hot Jupiters and have smaller orbits. Their sizes are around the size of the Earth, and their orbital periods are clustered around one day—ranging from as little as four hours up to a couple of days. (Note that there is a common definition for USPs as planets with periods less than one day, but that is a historical definition that arose because it was the cutoff for the original Kepler analysis pipeline. It was not defined by some physical or statistical property of the systems that contained them.)

It takes some work to tease out the distinguishing characteristics of these hot-Earth, ultra-short-period planets. As we saw with hot Jupiters, the existence of a group of planets with a unique set of properties generally indicates a different formation or dynamical history. We know that the solar system didn't have the same history as a hot-Jupiter system, not only because there is no hot Jupiter orbiting the Sun, but also because the Earth is still here. Had Jupiter undergone high-eccentricity migration, the terrestrial planets in the solar system, like Earth or Venus, would not have survived. The eccentric Jupiter would have constantly crossed the orbits of the Earth or Venus precursors—scattering them into the Sun or ejecting them from the system.

Similar arguments about a unique formation history can also be made for systems with ultra-short-period planets—their properties indicate something from their past that distinguishes them from both the solar system and the more typical planetary systems that Kepler discovered. Early work on these planets came from a comparison of systems with one transiting planet candidate (singles) versus systems with multiple planet candidates (multis). It was clear by the second year of Kepler data that single- and multiplanet systems showed subtle differences in the planet sizes and orbits. Single planet systems often had larger planets when compared with multiplanet systems, and the orbits of those planets were preferentially shorter (such as one or three days).

It's important to remember that just because Kepler only detects a single planet in a system, it doesn't mean that there are no other planets in that system. If the planetary orbits are slightly misaligned, even by less than one degree, it is enough for some planets to be undetectable.

It's possible, perhaps likely, that single-planet systems are simply multiplanet systems where the other planets don't happen to transit. By comparing the statistical properties of planets in the single-transiting and multitransiting systems, such as sizes and orbital periods, you can find populations of planets that appear in one group but not the other. We have already seen one population of planets that appear in single-planet systems that do not show up in multiplanet systems: the group of Jupiter-sized planets with three-day orbital periods.

A few years into the Kepler mission, I ran another study comparing single- and multitransiting systems, this one involving more planets than the original study (thousands of planets instead of hundreds). We showed that not only were hot Jupiters different from the typical planetary system seen by Kepler, but the Earth-sized ultra-short-period planets were also a unique population of exoplanets. As with the hot Jupiters, our result implied that the formation and dynamical histories of hot Earths also differ from most other planets. In other words, it was almost impossible for the physical processes that produced the sizes and orbits of most Kepler planetary systems to simultaneously produce the observed number of hot Earths. Something was happening where either the conditions caused planets to form right next to their host star, or they formed farther afield and some other process brought them closer and deposited them on these one-day orbits.

One possible origin story appeared in a study led by Francesca Valsecchi at Northwestern University (I was a coauthor on this work). We wondered whether these hot Earths might be leftover cores of hot-Jupiter planets. In this scenario, the hot Jupiter gets too close to the host star, and the star's gravitational pull exceeds that of the planet—causing the star to strip the atmosphere from the orbiting planet through a process called Roche lobe overflow. Surviving this process would be the rocky material from the core of the planet. We found that this process was indeed a viable explanation—it could happen. If it turned out to be common, it would imply that the number of hot-Jupiter systems was originally much larger (two or three times larger) than what we see today, but half to two-thirds of the original hot Jupiters would have disappeared over time as their atmospheres were consumed [39].

The idea of hot Earths being stripped cores of gas giants was a good idea to consider and to test. One test is to see whether the other properties of hot-Earth systems match the properties of hot-Jupiter systems. If these planets were indeed leftovers from adventurous hot Jupiters, then the stars should look the same between the two groups. Unfortunately for us, it turns out that the properties of the host stars of hot Earths match more closely the stars of typical Kepler planetary systems, rather than the properties of systems with hot Jupiters. Hot-Earth systems didn't show the high abundances of heavier elements in the stellar atmospheres that we see in hot-Jupiter systems. So our hypothesis, as cool as it was, may contribute a little to this population of hot Earths, but it cannot be the primary means to form them.

A clue to their origins might be found in the fact that many hot Earths appear in multiplanet systems. The Kepler-10 system is a good example. While the rocky inner planet Kepler-10b made the headlines, there is a second planet much farther out—Kepler-10c, with a six-and-a-half-week orbit. A six-week orbit may not sound like much (it is about half the orbital period of Mercury), but when compared to the twenty-hour orbital period of the inner planet, Kepler-10c may as well be orbiting a completely different star. In terms of their dynamical interactions, these two planets are incredibly far apart. To visualize this, imagine a rearrangement of the solar system. If you remove all of the planets from the solar system except for Venus and Saturn, keep Venus the way it is, and remove eighty percent of the mass of Saturn, giving it the mass of Neptune, you get the equivalent of the Kepler-10 system. Venus and a small Saturn, separated by a huge gulf of nothingness, is how dynamically isolated the inner planet of Kepler-10 is from the outer planet.

Many hot-Earth systems have a similar orbital "architecture" to Kepler-10. That is, they have a close-in, Earth-sized planet with one or more sibling planets on much larger orbits. It turns out that, in these systems, the closer the hot Earth is to the central star, the more its orbit becomes detached from those of its siblings (measured in terms of the ratio of the orbital periods of the hot Earth and the next-nearest planet). The planets that are farther from the central star, with orbital periods beyond a few days, can be quite close together, with the periods of adjacent planets often differing by as little as twenty percent. But, when we see systems where the planets get very close to the host star, the innermost planet seems to pull away from the rest of the family—as though it was dragged toward the central star, leaving the other planets behind.

In addition to this separation, the orbits of these hot Earths appear misaligned with the orbits of their siblings. In these systems, we often find several outer planets whose orbits all align to within a few degrees. Yet the orbit of the innermost planet is off kilter by 10 or 15 degrees, sometimes more. This is a huge difference by solar-system standards, where the orbital planes of the planets are within a few degrees of each other. You have to go out to the non-planet Pluto to see orbits that are as misaligned as the hot Earths in these systems. In fact, the only reason we can see the inner hot Earths transiting in these systems is because they are so close to the central star that even large misalignments can still have the planet cross our line of sight. The orbital architectures of hot-Earth systems are similar to a scaled-down version of some hot-Jupiter systems—those that have exterior giant planets on wide orbits. This fact suggests that the hot Earths may also have formed from eccentric migration within these systems—possibly driven by these exterior planets.

The observed orbital misalignments mean that for each discovery of a hot Earth that is part of a multitransiting system, there are several more where we see only the hot Earth, or only the other planets, but not both. If you correct for the rarity of seeing both the outer and the inner planets, then you can account for nearly all the hot Earths that we observe. That is, hot Earths are almost all members of multiplanet systems where their orbits are inclined from their neighbors, and they aren't therefore stripped cores of hot Jupiters or Neptunes. It is possible that the isolation of these planets is a consequence of their orbital eccentricity being excited by their previously nearby planetary siblings, and then the tidal force from the star acting on the planet, circularizing its orbit and pulling it inward. Or it may be that they are a result of interactions with the inner edge of the planet-forming disk.

But, whatever dynamical history put them in their predicament, it was certainly a different history from the solar system.

Given the differences in the histories of these planets, let's take a moment to compare hot-Earth exoplanets to their nearest counterpart in the solar system. The hottest planet in the solar system is ... well, the hottest planet is Venus, because of its massive greenhouse effect, which is not what I was going for. Rather, let's compare them to the planet nearest the Sun, Mercury. Mercury orbits the Sun every three months, rather than every few hours. It orbits at a distance twenty times larger than these hot Earths. Despite the larger orbital distance, Mercury participates in many of the same interactions with the Sun that these planets have with their host stars. We've noted that the tidal force from the central star distorts the shape of the orbiting planet, and internal friction causes the planet interior to heat up. The energy to heat the planet comes in large part from the planetary orbit, but it can also come from the planetary spin.

When energy is drained from the orbit, the orbit shrinks. When energy is drained from the spin, the spin slows down. Over time, the internal friction from the tidal interaction with a star will cause a planet's spin to slow until it couples to the planet's orbital period. If a planet spins at the same rate that it orbits, it will keep the same face toward the star at all times. In this situation, the distortion to its shape from the tide—which always points toward the star—will stay fixed relative to the star and the internal friction dies away. The planet settles into a 1:1 spin-orbit coupling, where neither its spin nor its orbit loses any more energy to friction-caused heating of the planet interior. It is this process, for example, that drove the Moon to its current spin state, where it always keeps the same face toward the Earth. In the past, the Moon spun much faster on its axis, but over the intervening billions of years, internal friction caused its rotation to slow until it matched its orbit—which is the state where we see it now.

A similar thing happened to Mercury in relation to the Sun, with a slight twist. Mercury has a more eccentric orbit than the Moon. When Mercury is close to the Sun, it sweeps around the Sun at a rate that is faster than when it is far from the Sun. If Mercury actually settled into a 1:1 coupling between its spin and its orbital period, like the Moon, it would mean that when it is close to the Sun, it would orbit faster than it spins for a brief period. Then, when it was farther from the Sun and its orbit slowed, the reverse would happen and it would spin faster than it orbits. This ever-changing difference between its spin and orbital rates would, again, cause friction in the interior of the planet—meaning that the system would continue to evolve until the cause of that friction was removed.

Under most circumstances, this constant source of friction would circularize Mercury's orbit. But in this case, as the shape and size of its orbit evolved, it settled into a different relationship between its spin and orbit—a 3:2 coupling. Today, when Mercury passes close to the Sun, its orbit and spin rates nearly match. Then, when Mercury moves away from the Sun, it maintains that spin rate, while the orbit slows. After Mercury has traveled one-third of the way around its orbit, it will have spun halfway around on its axis. Another third of the way around its orbit and it will complete one full spin revolution. At the end of a single orbit, it will have rotated one and a half times, with the original dark side of Mercury now facing the Sun and the original sunlit side in the shadow. The process continues for another orbit, Mercury spins one and a half more times, and now it returns to its initial state with the original daylight side back on the daylight side again—with two orbits around the Sun and three complete rotations about its axis.

Mercury's proximity to the Sun and relatively small mass mean it has no substantial atmosphere. (It does have one. But it is one hundred-trillionth of the Earth's, and is largely made from particles stolen from the passing solar wind.) The lack of an atmosphere prevents heat from transferring from the daylight side of Mercury to the nightside. As a consequence, the dayside has a temperature near 800°F, while the nightside is nearly −300°F. The coupling of the spin and orbit add another complexity to Mercury's surface temperature. Some parts of Mercury will never directly face the intense heat from the Sun during one of Mercury's close approaches. Instead, if you divide Mercury into four orange slices, each approach will bring one pair of opposing slices directly beneath and directly opposite the Sun. These two slices trade places from one orbit to the next. The other two slices will always be at sunrise or sunset during successive close approaches. Thus, if you remove the day–night variation in Mercury's surface temperature, and look at the average temperatures across Mercury's surface, it is divided into four zones—two "hot" and two "cold," each spanning a quarter of the planet.

Hot-Earth exoplanets are subject to the same physics as Mercury, with a related result. Because their orbits are circular, they will bypass the 3:2 spin-orbit coupling that Mercury experiences, and will head straight to a 1:1 coupling like the Moon has with the Earth. These planets will have one side that is constantly facing the star and the other side constantly pointing away. While the ones we've seen are more massive than Mercury, their closer proximities to their host star mean that their atmospheres, if they have one at all, will be similarly thin, with a similarly inept ability to convey heat from the perennial dayside to the perennial nightside of the planet.

The dayside, ten times closer to its star than Mercury is to the Sun, receives a hundred times the energy for each square meter of its surface. Temperatures can reach thousands of degrees, melting the surface rocks. In some cases, as we saw with KIC-1255, it can vaporize the rock entirely, producing a tail that trails behind the planet as it orbits. Meanwhile, the nightside temperatures will be well below zero—with no chance to warm. One may think that the heat could transfer through the surface material if given enough time, but rock is a poor heat conductor and excess energy near the surface will radiate away before it is able to creep all the way around to the antistellar point.

That is the situation for rocky planets with a solid surface. The case is different for hot Jupiters that are made mostly of gas. There, the thick atmospheres can actually deliver energy from one side of the planet to the other. Like their hot-Earth counterparts, the average rotation of hot Jupiters on circular orbits will also be tidally locked into a 1:1 relationship. But the atmospheres of planets tend to have minds of their own. A rotating fluid will develop jet streams like what we see on the Earth, and like what we see in the banded colors of Jupiter's clouds. The jet streams that form in this atmosphere can carry recently warmed air past the point directly under the stellar gaze and toward the nightside—shifting the hottest part of the planet slightly downwind.

We can observe this shifted hot spot in a couple of ways. One way to see it is by looking at the light coming from the planet throughout its orbit (its phase curve). If the planet has a warm spot that is downwind from the substellar point, then the planet will appear brightest at a slightly different part of its orbit than it would if there were no wind. Imagine if the right-hand side of the full Moon was brighter than the left-hand side (which it already is since the heavily cratered highlands are lighter in color than the darker maria). Then, as the Moon goes through its phases, when the lighter side is visible, the Moon will appear brighter than when the darker side is visible. This effect is most exaggerated during the crescent or quarter phase when the exposed side is either primarily the bright highlands or the darker maria. You would be able to tell that the Moon's surface had different reflecting properties just because otherwise identically sized crescent phases (the first one just before the new Moon and the second one just after) would have different brightnesses.

The same offset in brightness can happen with the phase curve of a hot Jupiter. If there were no wind, and the hottest spot was directly beneath the star, the phase curve would be symmetrical. But if the wind circulates warmer air around the planet, and the hot spot is shifted from directly beneath the star, then the phase curve will no longer be symmetrical. It will be brighter when the hot spot is exposed and dimmer when it is hidden.

A second way to see the effects of winds on a hot Jupiter is to look at the *secondary eclipse* of the planet—when the planet passes behind the star. During the secondary eclipse, the light that the planet reflects or emits is blocked by the star. Here, again, if there are brighter and darker halves of the planet, they will show up in how the overall brightness of the system changes as the planet hides behind the star, and when it reappears. If the darker portion of the planet gets blocked first as it passes behind the star, the system will dim more slowly than if the lighter portion is blocked first. The inverse will happen when the planet reemerges from behind the star. If the light portion of the planet appears first, the whole system will brighten more quickly than if the darker portion appears.

This *eclipse mapping* approach was used to produce the first map of the surface of an exoplanet. In this case, the planet was HD 189733 [40]. The data were taken in the infrared, where the light coming from the planet is brightest. The researchers, led by Heather Knutson (then at Harvard, now at Caltech), used observations of the planet taken by the Spitzer Space Telescope. Part of that team was my PhD advisor, Eric Agol, who always seems to find his way into the most interesting topics in astrophysics, as well as Eric's second student (my academic younger brother), Nick Cowan, who is now at McGill University in Canada.

Heather and her group found that, indeed, the winds in the atmosphere of HD 189733 blew the warm air eastward from the substellar point—in the direction of the planet's rotation. Subsequent studies of other systems show that they too have similar displacements of the warmest part of the planetary surface. (I'm using the term "surface" loosely here, referring to the layer in the atmosphere where the light originates. These are gas-giant planets with no real surface after all.) Only a small number of planets have a bright spot that lags behind the planet rotation, including CoRoT-2 and Kepler-7. The signals from these two weird planets may be the result of clouds, or (in the case of CoRoT-2, which is still quite young) the planet may not yet be tidally locked, and therefore may still rotate more slowly than it orbits, which would make the winds appear to blow in the opposite direction from the substellar point.

Continuing the comparison of hot Jupiters and hot Earths, the process of shrinking and circularizing the planetary orbits through tidal friction dumps a lot of energy into heating up the planetary interiors. This doesn't make much difference for rocky planets with no atmospheres. Rocks don't change their size much when they heat up or cool down. This is not to say that rocky planets don't change at all as they cool: We have expansion joints in bridges on the Earth so that they don't buckle when the temperature changes. And we know that when

Mercury's core cooled and shrank, the crust fractured and collapsed in places, leaving behind large cliffs hundreds of miles long, called *scarps*. But extreme temperatures on rocky exoplanets would cause their sizes to change by an amount far below our ability to detect.

Gases, on the other hand, expand much more readily as their temperature rises. The combination of intense heat from the nearby star and energy drawn by the atmosphere from the shrinking and circularizing planetary orbit causes the atmospheres of hot Jupiters to inflate to substantial sizes—often several times larger than Jupiter, despite having similar masses. The volumes of these inflated exoplanets can be nearly ten times as large as they would otherwise be. The result is some gas-giant planets with surprisingly small densities. A rocky core, perhaps, but with enormous fluffy atmospheres.

For comparison, Saturn is less dense than water. We tell schoolchildren that if you filled a giant bathtub with water, Saturn would float. (Of course, this visualization requires not only a bathtub a thousand times more voluminous than the Earth, but also a gravitational field in which the water can rest so that the planet can float on it, but those are merely details. The fact remains that Saturn is less dense than water.) Water's density is one gram per cubic centimeter. Jupiter has a density thirty percent larger than water. Saturn's density is about thirty percent less than water. Jupiter would sink. Saturn would float, with thirty percent of its volume protruding above the water line. Some inflated hot Jupiters have densities about one-tenth that of water. Indeed, when the first batch of Kepler planets was announced, Kepler-7 was dubbed the styrofoam planet since its density was similar to some types of styrofoam (or popped popcorn if you prefer).

When the density of Kepler-7 was first measured, the members of the team were alarmed at how small it was. We were concerned that there was something wrong with our observations. Not seeing any error, we moved forward with the announcement. Over time, more fluffy planets turned up, and our fears were put to rest. Nevertheless, it is still strange to think of planets so light. As more measurements of planet sizes (and hence, densities) were made by astronomers, it appeared that the lower the mass of the hot Jupiter, the more likely their atmospheres would be inflated. In other words, the lower the planet mass, the less gravity it has to keep its atmosphere under control.

This trend of low-mass, high-fluffiness planets holds true as you observe less and less massive planets until, eventually, the planets give up trying to retain their large, gaseous envelopes. In the data, we've seen hot Jupiters and hot Earths, but absent from the Kepler data is any substantial group of hot, Neptune-sized planets. The primary reason we haven't talked about hot Neptunes so far in this chapter is that there aren't any. Or, rather, there are exceedingly few. There is at least one planet (not found by Kepler) that is Neptune sized on a roughly one-day orbit, but it is likely to be an example of a rare planetary core from a gas giant that has been stripped of its atmosphere [41]. The lack of hot Neptunes, or the hot-Neptune *desert*, is fairly easy to see in both the Kepler data, and in data from other transit surveys. Given Kepler's sensitivity, if they did exist in large numbers, we would have found them.

Here, when I say hot Neptunes, I'm referring to Neptune-sized planets specifically. The physical properties of the Neptune-sized exoplanets that Kepler does see may not match the properties of their solar-system namesake, because of where and how they formed. Planets like Neptune and Uranus are different from both the terrestrial planets like Earth, which are made almost entirely of metal and rock, and gas-giant planets like Jupiter and Saturn, which are made mostly of hydrogen and helium. Instead, the solar-system *ice giants* are essentially a solid core of a few Earth masses, surrounded by a thick ocean of light molecules such as water, methane, and ammonia. This is the composition that arises out in the hinterlands of the solar system, because the temperatures are cold enough for the light, icy compounds to condense and coagulate into a planet. A similar process happened with the Earth and its neighbors, except that with the Earth it was the rocky material that condensed. (Gas-giant planets don't form this way, as we've seen. While they may start from a solid core, once they get going they grow primarily by stealing gas directly from the protoplanetary disk.)

Neptune-sized planets in the Kepler data, even if they don't orbit as closely as the hot Jupiters and hot Earths, still orbit much closer to their star than either Neptune or Uranus orbit the Sun. Thus, Neptune-sized exoplanets may not be made of the same collection of ices. Rather, they may simply be a solid core with an atmosphere of hydrogen and helium. We don't know for sure, and there may be some variation in their composition. In the coming years we'll have better information about what their atmospheres are made of. For now, this core-plus-atmosphere model fits the data we have.

An atmosphere of hydrogen and helium may be why we see lots of roughly Neptune-sized planets in Kepler data, but only a small handful with the short orbits of the hot Jupiters and hot Earths. They may not be able to keep hold of their atmospheres if they are that close to the star. Jupiters are massive enough to retain their atmospheres under these conditions, but if you reduce that mass by two-thirds or more, the planet starts to lose ground to the other forces acting upon it. The combination of intense radiation from the central star, leftover heat from their formation, heat generated by tides as their orbits are shortened and circularized, and their smaller mass compared to their gas-giant counterparts, may simply boil off the atmosphere of erstwhile hot Neptunes and leave behind the naked core. If the atmospheres of hot Neptunes are boiled off the planets, the hot-Earth population of planets would be a combination of true Earth-sized planets and former Neptune-sized planets with the atmosphere stripped off. Like most questions in exoplanet science, there is still active research to try to understand why there are so few hot Neptunes.

Before Kepler, we knew about hot Jupiters. They appeared in both Doppler surveys and transit surveys. We even saw the phase curves and reflected light from hot Jupiters prior to Kepler. But Kepler's high-quality data teased out a number of additional effects on these systems, such as Doppler beaming and the planet-induced distortions to the host star (ellipsoidal variations). Kepler data also allowed us to survey the neighborhoods of these planets, looking for—but generally failing to find—nearby smaller planets that would be trapped there under the disk-migration model of hot-Jupiter origins.

By contrast, hot Earths were a discovery that was largely unique to Kepler. The lack of hot Neptunes in either transit or Doppler searches was still unrecognized until the Kepler data made their absence glaringly clear. (Before Kepler, we had the sensitivity to detect hot Neptunes, but there wasn't much chatter about their absence. The silence of the hot Neptunes was deafening in the Kepler data, where both larger and smaller planets were found in abundance.) Before Kepler, the only common disintegrating objects were solar-system comets made mostly of ices. An evaporating rocky planet wasn't a possibility that regularly surfaced in scientific discussions. All of these hot planets brought new information to the table when it comes to our understanding of planets and planet formation—settling some questions, and raising others. It also made clear how fortunate we are that our system does not have one.

5

Planetary Dynamics

Leading up to our meeting in Denmark and the first catalog release of Kepler data in June 2010 (the one with the four hundred sequestered planet candidates), an issue arose with how to treat systems with multiple transiting planet candidates. We knew that the Kepler data analysis pipeline would miss some planet candidates—no mixture of computer algorithm and human attention would be foolproof. Because of this, we knew that there would be multitransiting systems among the three-hundred-plus planet candidates that were included in the catalog. Rather than saying nothing, or pretending that we didn't find any such systems, we decided to include five systems that showed multiple transiting planets in the catalog and to write a specialized paper introducing them [42].

At the time, these were still planet candidates, not yet vetted as real planets. Since then, with more data and more scrutiny, they have all been verified as planets—along with some additional planets in these systems that were missed in the first bit of Kepler data.[1] There was a fair amount of discussion about which planets would be included in this work and why. One system we considered had already received a lot of attention from our group because of the orbital period changes we saw in the planet candidates—the future Kepler-9. Another possible system already had four planet candidates in just the engineering data and the

1. These are the now four-planet Kepler-79 system, the now three-planet Kepler-81 system, the still two-planet Kepler-117 system, the still two-planet Kepler-248 system, and the now three-planet Kepler-487 system.

first quarter. Because the number of planets in this system was so large, we decided to gather additional data and study it in more detail before making it public (the eventual six-planet Kepler-11 system).

Multitransiting systems were a major boon to understanding the properties of whole systems of planets, not just individual ones. Their value goes beyond just boosting the numbers of detected planet candidates. For example, through Doppler surveys, we had known about systems with multiple planets for years, but none of those previous discoveries came from transiting systems. Consequently, we had no idea of the sizes of the planets, and we had only limited estimates of their masses since, in order to measure the mass of a planet, we need to know the orientation of the planetary orbit. This is easy to accomplish if the planet transits, but quite difficult if it does not. With multiple transiting planets we can measure relationships among the planets that we can't get by other means, such as correlations between their sizes or their orbital properties. These measurements, in turn, give information about the formation and dynamical evolution of each system or collection of systems.

Relationships between planetary orbits in a system are a powerful window into the dynamics and the history of those systems. We've already seen from Le Verrier that the gravitational influence of one planet can distort the shape of another planet's orbit. That fact led to the prediction of the existence of Neptune. We can do the same thing with exoplanetary systems—gravitational perturbations to the orbits show up as TTVs—transit timing variations, or deviations from constant intervals between successive transit times.

Some TTV signals occurred in systems with only one transiting planet, while others were in systems with multiple transiting planets. Both kinds of systems are interesting for different reasons, so this left us needing to choose which system or type of system to study first. The advantage of studying a system with a single planet showing TTVs is that we could use them to find and characterize the as-yet-unseen planet that was causing the perturbations. This would be a reapplication of Le Verrier's method, nearly two centuries later and a thousand light-years distant. (Think about that. Because of the time it takes

for light to travel to us, we're applying Le Verrier's method to planetary perturbations that happened a thousand years before Le Verrier
was born.)

If we see a TTV signal in a single-transiting system, there can be
several locations where a perturbing planet can produce something
similar to the observed signal. It's a challenge to pinpoint all of the
essential properties of the unseen, perturbing planet to claim a new discovery, such as constraining its mass and identifying the correct orbit.
It would take more data to do this than what Kepler currently offered.
By contrast, with multiple planets transiting in the same system, a lot
of the ambiguity is removed because we can see which planets cause
the perturbations. In fact, because we know from the transit signals
what the orbital periods are for each of the planets, it is much easier
to see the effects that each planet has on all of the others. With this
extra information, estimates of the properties of the planetary orbits
and their masses are easier to make. More importantly, for this stage in
the mission, we could make these measurements with the data we had
in hand—not needing to wait for more information to come down from
on high.

There were two of us who joined the Kepler team specifically to
do this kind of analysis: myself and Matt Holman from Harvard. In
2005, we coordinated the simultaneous appearance of our two papers
that showed the potential power of transit timing variations. There had
been a few early searches for TTVs (I had conducted two searches
already), but all of them had come up empty handed. Now, we both
wanted to lead the effort to analyze the first definitive TTV signal.
We knew it would be a major breakthrough in exoplanet science. This
could have been a major point of contention within our group, which
we both wanted to avoid. In a late-night, late-winter 2010 phone call
between Matt and me, we agreed that Matt would lead the study of
the TTV signal in this multiplanet system, I would continue to lead
the paper announcing the first five multitransiting systems, and later,
with more data in hand, I would lead a paper on the first detection of a
non-transiting planet with this same technique—a TTV signal from a
single-transiting system.

It turned out that nature had other plans for the first detection of a non-transiting planet through TTVs. While we were waiting for a good candidate system to materialize, Sarah Ballard (a graduate student from Harvard, now faculty at the University of Florida) was studying the transits of what became the Kepler-19 system. She was using infrared observations from the Spitzer Space Telescope to verify that the infrared properties of the transit signal were consistent with a planet instead of a small star—small stars emit lots of infrared light, while planets generally don't, so the transit shape in the infrared is different for stars and planets. In a spring 2011 meeting of the science team, she presented her work, which included a weak but unmistakable TTV signal.

After a lengthy, somewhat uncomfortable discussion about what to do to try to honor the scientific interests and commitments to the different individuals (Sarah and me) and groups (the Multibody Working Group and the group working with Spitzer data) within the science team, we all realized that there was nothing that could reasonably be done to salvage either of our original plans. We couldn't wish away the presence of the second planet in Sarah's system—it was there whether we liked it or not. Her work was important, so we couldn't have it languish for months while we waited for more data on a different system to arrive and for its analysis to be completed. Chance had placed the science team, and two of its youngest participants, in a difficult spot. It was not an easy predicament to navigate, and was an unusual start to our professional friendship, but the team ultimately made the right call to move forward with Sarah's work, and the Multibody Group assisted in interpreting its TTV signal.

But that situation was still a few months in the future. With the agreement hammered out between Matt and me, our group started working on the analysis of KOI-377—the eventual Kepler-9 system. This involved studying the dynamics of how the planets in the system interacted and affected each other's orbits. It also included a suite of ground-based follow-up observations that were needed to show that Kepler-9 was not some astrophysical false positive. From the dynamical analysis, we predicted when the system would show a large Doppler signal. Turning that information over to the Kepler Follow-up Observing

Program, they made a handful of Doppler observations with the Keck telescope. Their measurements confirmed the dynamical prediction and gave good constraints on the masses of the two planets.

The studies of the Kepler-9 system and the five multitransiting systems competed for the time and resources of our group. Our working group was so excited by Kepler-9 that studying this one system consumed more of our collective attention than the combined projects of the first release of the Kepler catalog and the announcement of the detections of multitransiting planetary systems stated above. Initially, I had taken over the leadership of the paper with five multitransiting systems, because it needed to get done, and no one else was doing it—Kepler-9 was virtually all-consuming. The late-night phone call solidified our respective roles in bringing those results into the scientific literature. Initially, we hoped to publish both Kepler-9 and the five multitransiting systems in conjunction with the first Kepler catalog, but time was short and the desire to be thorough made the Kepler-9 result slip down the calendar.

Part way through the Kepler-9 analysis, one of the scientists (Darin Ragozzine, who was a postdoctoral scholar and is currently a professor at Brigham Young University) thought he saw something else in the data. He had run an additional transit search, just to make sure that we hadn't missed anything. What turned up was a small, roughly Earth-sized transit-looking signal with roughly a one-day period—another planet candidate orbiting this star. He described Kepler-9 as the system that kept on giving; I thought of it as the system whose analysis would never end because it kept pulling time and attention away from my project.

While we missed our target for publishing Kepler-9 along with the first catalog in June 2010, it came out three months later in October [43]. We submitted that work to *Science*, one of the world's most prestigious scientific journals. For full disclosure, at the time, I advocated against trying to get it into *Science*, thinking it would slow the paper down. Looking back, the three-month delay wasn't a big deal, and I'm glad that Matt stuck to his guns and followed through with his intention to publish there. His patience working it through the system paid off. When the study did arrive, it was the cover story for the magazine, with an inspiring artistic impression of what the system would look like close up. (The original concept art for the cover had a semitransparent overlay of the footprint of the Kepler camera on the planetary image, which I thought was even better than what actually appeared on the cover.) Plate 7 shows an artist's rendition of the Kepler-9 system.

Kepler-9 was the first confirmed multiplanet system from Kepler (as opposed to systems with planet candidates). The orbital periods of the two planets are twenty days and forty days. This 2:1 orbital period ratio likely does not represent the orbital periods that the planets had as they were forming, and is likely the consequence of the two planets moving relative to each other. We saw this in the solar system with the 3:2 period ratio of Neptune and Pluto, caused by Neptune's outward migration into the Kuiper belt of icy debris in the outer solar system. We suspect that a similar migration-plus-capture scenario played out in Kepler-9 to place the two planets near the 2:1 mean-motion resonance.

This orbital configuration, with planetary periods like 2:1 that are near ratios of integers, produces a large TTV signal. Whenever two planets pass each other in their orbits, when they have a *conjunction*, they exert a small gravitational force on each other. (Here, the force is small compared to their interaction with the central star.) If the two planetary orbital periods have no relationship with each other, then the conjunctions happen at different locations around their orbits and their effects will offset each other over time. Sometimes the conjunction will cause a planet's orbit to grow slightly, and sometimes it will shrink it a bit. However, when the two planets have orbital periods that are the ratios of integers, like what we see with Kepler-9, the conjunctions occur near the same part of their orbits. Now, rather than quickly canceling each other out, the small gravitational effects add. The compounding perturbations cause the orbital periods of both planets to change: when one planet's period gets longer, the other gets shorter, and vice versa. These changes, in turn, will alter the location where the conjunctions take place. Eventually, the shift in the location of the conjunction will cause the effects of the perturbations to slow and reverse. The result is that the two planet's orbital periods oscillate (growing larger, then smaller, then larger again) over a long-term *TTV cycle*.

These changes can be quite dramatic. In Kepler-9, the orbital periods change by almost a day throughout their TTV cycle. Keep in mind that the planets in Kepler-9 only have orbital periods of about a month, so this constitutes a change of several percent in the orbital period of the planets. If the Earth went through similar changes to its orbit, we would have to add two weeks to the calendar over the course of one decade, and then take them off again during the next decade. Other planets in the Kepler data have TTV signals that are even larger, changing by ten percent every few orbits—the equivalent of adding, then later subtracting, a whole month every few years.

The two primary planets in this system are similar in size, and their orbits are nearly (though not quite) circular. All of this information tells us what happened, or what didn't happen, in their past. Having these near-resonant planet pairs tells us that the planets likely migrated toward each other at some point—possibly while the gas disk or a disk of planetesimals was still around. They did not undergo any violent scattering events that would distort the shapes of their orbits into ellipses.

Following closely on the heels of the Kepler-9 discovery, another system consumed the attention of the Multibody Working Group. Initially the KOI-157 system (soon to be renamed Kepler-11) showed evidence for four transiting planets orbiting a single star. Over the subsequent months, additional data poured in from the spacecraft and another pair of planets materialized. By January 2011, with the second release of the Kepler data (this time with the full candidate list), the Kepler-11 system was announced—appearing as the cover story of *Nature*, another high-profile scientific journal [44]. Plate 8 shows an artist's rendition of the Kepler-11 planetary system.

The six planets in Kepler-11 orbit a star similar to the Sun, about four percent larger in mass and two percent larger in radius. All of these planets have orbital periods less than that of Venus, with five of the six orbiting interior to Mercury. We did a TTV analysis of the system—including all of the interactions of each planet with each other planet (though the sixth planet was considerably more distant from the other five and influenced them to a much lesser degree). This study placed the total mass of the inner five planets at roughly thirty times that of the Earth. Five planets, all interior to Mercury, with a combined mass thirty times the Earth—equivalent to the combined masses of Uranus and Neptune—had never been seen before.

Sure, there were other systems with a similar number of planets that had been detected in other ways, and the solar system is one example. But the orbits of the planets in these other systems span years, even decades—not weeks. In addition, the Kepler-11 system was incredibly flat, with the orbits of the planets all nearly aligned. By comparison, there is no vantage point from outside the solar system where you can get six of our eight planets to transit the Sun. Yet that was how things were in Kepler-11. The average difference in the orientation of the orbits in that system is about one degree, compared to the solar system where the planets are two and a half times further out of alignment, with Mercury being the most off kilter.

All of the planets in Kepler-11 are more massive, and larger in size than the Earth, while all but one is smaller than Neptune. All but one are less dense than Neptune, and they are all less massive than Neptune (the outermost planet may be an exception, but that is not likely). Jack Lissauer, who led the Kepler-11 study, described the Kepler-11 planets in the associated press conference as "fluffy . . . sorta like marshmallows. Maybe a marshmallow with a little hard candy at the core." The solar system has no examples of planets in this regime between Earth and Neptune (the two benchmark planets in their respective regions of the solar system).

Another unexpected property of the Kepler-11 system was how close together the planets were, especially the inner five planets. In the same press conference where Kepler-11 was announced, Jack indicated that Kepler-11 was "the most compact system of planets ever discovered by any technique anywhere." Kepler-11 was an important challenge to the theories of planet formation in all the ways we could measure its properties—the flatness of the system, the masses of the planets, and the compactness of their orbits. Kepler-11 was not like the solar system, and it was not like any planetary system we'd seen before. It broke the mold for what we could expect to find in our data, and we would eventually find a lot more.

The compact orbits were particularly interesting for our group. The two large planets in Kepler-9 were near the 2:1 mean-motion resonance. In Kepler-11, the orbital period ratios were all less than 2:1, with most less than 3:2, and the innermost pair (at ten days and thirteen days) near the ratio 5:4. These nearby, and near-resonant, orbits implied even more planetary migration than what Kepler-9 had shown. It may also indicate that the planets formed closer together than we otherwise might predict—it still isn't clear whether their close period ratios come primarily from the location where they formed, or the subsequent migration and dynamical evolution of the system.

The dynamics of a planetary system, meaning the orbital changes that occur through the interactions among the planets and the star, don't really depend on the actual orbital periods of the planets. Rather, they depend more strongly on the ratios of those periods. That is, the changes to the orbits of the Kepler-11 system, because of the gravitational interactions among the planets, would look almost identical if the planetary orbits were ten times larger than they currently are. An interaction might produce a one-percent change, but that would be a one-percent change regardless of the physical size of the system. (We measure time in years and days, but that is a human-centered system. The Earth is hundreds of light-years away from these planets, so their orbits are completely unrelated to how long it takes for the Earth to circle the Sun, or how long it takes to spin on its axis. A year or a day is a meaningless quantity for these planets, so quantities like the number of planetary orbits are more relevant.)

For Kepler-11, not only are the planets physically near each other, but dynamically speaking, they are incredibly close together. The inner two planets of Kepler-11, with an orbital period ratio near 5:4, showcase how compact the system is. If they were replaced with two Jupiter-mass planets, the system would be unstable—their gravitational effects on each other would drive them into each other, or drive one into the star, or drive one out into space. There are no planet pairs in the solar system that have period ratios this small. Neptune and the non-planet Pluto come closest at a ratio of 3:2. The next nearest pair in terms of period ratios is Venus and Earth which is near 8:5. To find orbiting bodies with period ratios as small as the inner two for Kepler-11 you need to get into moon systems of the giant planets, where the masses are sufficiently small relative to the planet they orbit that the perturbations don't cause catastrophe. Titan and Hyperion orbit Saturn with a period ratio near 4:3, and that is still more widely separated than these two planets in Kepler-11.

The two closely spaced planets in Kepler-11 aren't even the most extreme that we've seen. In December 2011, NASA held the first Kepler Science Conference at the NASA Ames Research Center. (This was the conference where I had the interesting exchange with a science journalist about the sequestered data from the first Kepler catalog.) At that conference, one system started making the rounds among the members of our Multibody Working Group and other exoplanet dynamicists during meals and coffee breaks. At the time, there was one planet candidate in the KOI-277 system, but its orbital period was erratic. Its period would stay constant for a few orbits, then it would shift to a different period and stay constant for a few orbits, then shift again. It was so weird that I originally blew it off as some sort of noise or artifact of the data analysis pipeline. After all, there were several other systems that had strange behavior because of noise or incorrect estimates of orbital periods, so I chalked this up as just one more of them. As time went on, however, the pattern kept repeating.

Since the behavior wasn't going away, a few of us started brainstorming ideas about what it could be. Maybe it was on an eccentric orbit, and the system had an outer planet with an orbit six or seven times longer. In that scenario, every several orbits there would be a particularly close encounter of the two planets which would mess with the orbit of the interior planet. Then the pair would avoid contact for another several orbits until the process repeated itself. Alternatively, perhaps this planet had a circular orbit and the distant perturber was on an eccentric one.

My graduate advisor (again), and a graduate student at Harvard named Josh Carter, spent some time on the system and found the cause of this behavior. Buried in the Kepler data was a small, hard-to-find signal of another transiting planet just interior to the known one. This interior planet's orbit had even larger deviations in its orbital period, which made it even harder to detect since the Kepler analysis pipeline assumes constant orbital periods in its search.

This planet pair has a period ratio near 7:6. So every 7 orbits of this new interior planet, and every 6 orbits of the exterior planet, they would have a conjunction. That conjunction would cause an abrupt change in each of the two orbits. Because the planets' orbital periods are so near each other, they orbit at similar speeds, so it takes quite a while for the inner one to lap the outer one. During this interval, the two planets continue at an approximately constant rate. Then the orbits are disrupted again, followed by a period of near-constant orbital periods. The average orbital periods of the two planets differ by only fifteen percent, and their orbital distances differ by only ten percent. At sixteen and fourteen days, their orbits are still short by solar-system standards, but by the standards of planets seen by Kepler, they are fairly typical. While the individual orbits are not unusual, the fact that they are right next to each other and the strength of their dynamical interactions are uncommon traits in planetary systems [45].

Despite the fact that their orbits are incredibly close together, the sizes and densities of these two planets are strikingly different. The smaller, inner planet is eight times more dense than the larger, outer planet. The outer planet is more massive by a factor of two, but it is two and a half times larger in radius—giving lots of room for the planetary material to fill. So you have two planets in the same system, practically orbiting on top of each other, with a density ratio equal to the density ratio between the Earth and Saturn, which is the most extreme pair that you can find in the solar system.

This Kepler-36 system is a challenge to form because of the difference in the planet densities and the close proximities of their orbits. If you tried to arbitrarily place a pair of planets into a system with a period ratio near 7:6, the most likely outcome would be the system destroying itself in a few orbits. In fact, most configurations near this period ratio lead to unstable systems, where one planet is either ejected or crashes into the star, or the two planets collide. For this pair, the planets likely formed both farther away from the star and farther apart from each other. Then, as we've seen with Kepler-9 and Kepler-11, they migrated inward, with the outer planet moving more quickly than the inner one and closing the gap between them. The issue is that, as the planets cross a period ratio of 4:3, or 5:4, stable orbits become increasingly rare. There is a minefield of system-destroying orbital configurations that the pair need to navigate as they converge—lest they no longer remain a system. Then, at some point, they need to abruptly stop their migration and stay where we see them today.

Clearly, systems like this can form, but it takes some finesse to make them work. This isn't the end of the weirdness that the Kepler-36 system shows. Matt Holman mentioned to one of his students (Katherine Deck) that it looked like Kepler-36 might be chaotic. Mathematical chaos is an interesting effect that often shows up in dynamical systems. Chaos boils down to the fact that the future of a system depends strongly—exceptionally strongly—on its starting point. Tiny changes to the starting conditions can give completely different outcomes down the road. This is the famous *butterfly effect*, where small disturbances like the flapping wings of a butterfly could eventually determine whether a hurricane will form on the far side of the Earth [46].

Chaos is a fairly common occurrence when you have multiple bodies interacting—two pendulums that are attached end to end will show chaos. In fact, a strong dependence on the initial conditions doesn't even need multiple bodies. For example, consider a solid pendulum— a pendulum that is suspended by a rod instead of a string. If you start the pendulum swinging with a small amplitude, it oscillates with a specific period. If you make a small change to that amplitude, the oscillation period will change by an almost imperceptibly tiny amount. Eventually, if you keep starting the pendulum with a larger and larger initial amplitude, you will start to notice that the period changes—it takes longer for the pendulum to swing from one side to the other.

However, it may take several oscillations for the difference to show up. For example, if the oscillation periods from two different starting amplitudes differ by one-millionth of a second, it would take a million oscillations before a one-second difference would show up.

If you take this to the limit and start the oscillation when the pendulum is upside down (with an amplitude of 180 degrees), the pendulum will stay at rest, pointed upward with an oscillation period that is infinitely long. Beginning from this extreme, if you make even a tiny change to its initial amplitude, it will start to swing. It may have really large swings, and it may take a while for it to get from one side to the other, but it won't take forever. Near this vertical configuration, a small change in the initial conditions is the difference between an infinite and a finite oscillation period—and differences don't get any bigger than infinity.

This strong dependence on the behavior of a system due to small changes in its starting position is the essence of chaotic motion. The solar system is filled with examples of chaotic motion—the future positions of all the planets have a similar dependence upon where they start. Often, however, those differences only show up over long timescales. The shortest timescale for chaos to appear in the behavior of solar-system planets occurs with the orbit of Mercury.

Because of the presence of the other planets in the solar system, the slightly squashed shape of the Sun, and a few other physical effects, the orbit of Mercury slowly changes its orientation relative to the Sun. It traces out a rosette pattern over time—like following the outline of the petals of a flower. The rate of this reorientation, or *precession*, is about one-sixth of a degree every century. The orbits of all other planets also trace out rosette patterns, and for the same reasons. The precession rates for the different planets are largely unrelated to each other. Earth is just under twice the rate of Mercury, Mars is over three times the rate, and Saturn is nearly four times the rate.

Jupiter, however, is another issue. It so happens that the precession rate of Jupiter is about one-fifth of a degree per century—slightly faster than Mercury, but only by about ten percent. This coincidence creates a problem. Mercury would like the orientation of its orbit to move at its own pace, but the influence of Jupiter tries to push it at a slightly different pace. Jupiter couldn't care less about the evolution of Mercury, and Mercury is able to ignore Jupiter most of the time. However, if Mercury were to increase its eccentricity by a little bit, that would speed up its precession rate to match Jupiter's more closely. It can't match Jupiter exactly, but it can get closer.

Occasionally, Mercury will happen to be in the wrong place (relative to everything else in the solar system) at the wrong time. Its eccentricity will start to grow, and its orbit will couple more strongly to the orbit of Jupiter—which can drive its eccentricity higher, coupling it more strongly, and driving its eccentricity even higher. Mercury can break out of this feedback loop, and usually does—returning to its preferred precession rate. For example, the increasing eccentricity starts to affect its spin–orbit coupling that we saw in the last chapter. The internal friction that arises as a consequence can damp the eccentricity back down. However, there is a chance that the growing eccentricity gets out of hand. Eventually, its eccentricity can get large enough that it might smash into the Sun, or its orbit might cross the path of Venus or the Earth, or it can disrupt Venus so that Venus smashes into the Earth. If you are reading this book, this catastrophic event hasn't happened yet, but it can. It just takes the accumulation of small changes in the orbits of the planets to put Mercury on this treacherous path, and the whole system to become unstable.

The potential instability of the orbit of Mercury takes a while to materialize. Orbital eccentricity doesn't grow like weeds in a garden. It would take hundreds of millions, or even billions of years with Mercury, starting where it is, to even have the possibility of ending up in the Pacific Ocean. The chances of these terrible circumstances arising depends very sensitively on the starting places of the solar-system planets. If we know exactly where the planets are, we can predict whether or not this will be a problem in the future. Right now, we know the positions of the planets to within a few meters. For the Moon we know the position to a centimeter or so. Is that good enough to tell whether or not we're all going to die in a large planetary collision in the future?

No.

The difference in the starting place today between a benign Mercury that leaves us alone and a Mercury that wreaks havoc in the inner solar system, and drives itself or another planet into a crash, is less than one millimeter. It is impossible to know the position of Mercury to within one millimeter, especially going into the future hundreds of millions of years. There are lots of ways that the orbit of Mercury, or the other planets, can be perturbed by less than a millimeter over the lifetime of the Sun. Even an asteroid, ejected from another star system, that sweeps through the solar system could cause a sufficient deflection to change the trajectory of the solar system a billion years from now.

The best we can do is estimate the probability of death by mercurial impact by examining the future of the solar system using a computer. By running a suite of simulations that begins with Mercury at a bunch of initial positions that are within our observational uncertainty, we can see what the collection of its future trajectories has in store. When this study was done in 2009, the French astronomers Jacques Laskar and Mickaël Gastineau found that the chances of the inner solar system going unstable are about one percent every billion years [47]. We are five billion years into the lifetime of the solar system, so we've been part of the lucky ninety-five percent of possibilities so far. We're now halfway to the finish line, at which point the Sun starts to expand and we will have bigger problems to worry about than Mercury.

With that happy thought in mind, let's look back at the Kepler-36 system. The pair of planets in Kepler-36 are near the 7:6 mean-motion resonance, but they are not actually in resonance. Being in a mean-motion resonance implies certain relationships in the evolution of the orbital properties of the planets (orbital properties like the precession rate and the rate of drift in the conjunctions due to the mismatch of the orbital periods from the exact numerical ratio). For Kepler-36, those relationships are not satisfied for the 7:6 period ratio. However, those relationships are satisfied for two separate mean-motion resonances near a different ratio of orbital periods, 34:29. When you account for the drifting in the orientation of one of their orbits (the precession rate), every 34 orbits of the inner planet coincides with 29 orbits of the outer one.

The problem is that the orbits of both planets precess, tracing out their separate rosette patterns, and they do so at approximately the right rate. If this were the only interaction that the two planets experienced, the pair would choose one of the options, picking which one of the two competing precession rates to follow and which one to ignore. However, the planets do experience other interactions, namely the very strong conjunctions that occur roughly every seven orbits of the inner planet and six orbits of the outer planet. The strong encounters that the two planets experience because their orbits are so close together cause the pair to occasionally switch between the two available resonances. They switch which precession rate they follow with their orbits, and which one they ignore. The rate they end up with after each encounter depends very sensitively on the initial conditions. In other words, it is chaotic.

After thinking about this for several hours, here is the best analogy I could come up with—despite its flaws. Suppose you go out for a run along a path that periodically crosses a series of small streams where you need to take care with your footing as you hop across to the other side. Naturally, as you run you listen to rock music from the 1980s, because there really isn't any other music worth listening to in these circumstances. Your fond memories of marching with the drumline in high school forces (at least initially) your footfalls to land in time with the music. Your left foot initially landing on beats one and three of each measure, and your right foot landing on two and four.

When you arrive at a stream, you break your cadence for a bit so that you can cross safely—you don't want to fall into the stream. Arriving at the other side you resynchronize with the music. Not wanting to break your general stride, your footfalls may now land on different beats within the measure. Sometimes you sync back up with the left foot on one and three, and sometimes you sync back up with the left foot on two and four. Since these stream crossings come up every few minutes, throughout your run your feet will be switching multiple times relative to the song. The foot you arrive on after each crossing might appear random, but it really isn't. It is sensitive to the width of the stream, the pace of your run, where the bank of the stream starts in your stride, and so forth. It is sensitive to the initial conditions, and the result is chaotic behavior.

This is essentially what happens in the Kepler-36 system. It has two ways for the planets to sync up their orbits, one from each of two possible 34:29 resonances. But roughly every six or seven orbits the planets have their regular conjunction—this is like when the jogger reaches a crossing. The effects of those conjunctions can cause the system to switch from one resonance to the other. (In some cases, the pair can be knocked slightly out of both resonances for a period of time.) Whether the system makes the switch or not depends upon the initial conditions, the exact locations that they have in their orbits, the details of the shape of their orbit at the time of conjunction, etc. Tiny differences can affect the results far into the future, and that sensitivity to the conditions of the system is the source of the chaos. Two seemingly identical systems end up in very different states at a later time because of the compounding differences that arise along the way.

For these kinds of systems, we characterize the chaos by measuring the Lyapunov time, or the time it takes for two nearby paths to diverge by some factor. The Lyapunov time for Mercury is just over one million years, a little more than five million Mercury orbits. This means that we would be able to distinguish between Mercury's current orbit, and a similar orbit where it was displaced by a millimeter from its current position, in about one million years. We've seen that after many Lyapunov timescales, Mercury's orbit can be destructively different from what we see today.

For the Earth, the Lyapunov time is just under five million years— tiny displacements in the Earth's orbit wouldn't show up in any meaningful way for about five million years. Jupiter's is nearly ten million years (roughly one million Jupiter orbits). The Lyapunov time for Kepler-36, on the other hand, is ten years. That's only two hundred and fifty orbits. That is an incredibly short amount of time for a long-term dynamical process in a planetary system. Over the lifetime of the Sun, Mercury will go through about ten thousand Lyapunov timescales. For Kepler-36, it will have experienced a billion of them. Kepler-36 today has traversed a longer, and possibly richer dynamical history than the solar system will ever experience. It is systems like Kepler-36, with all its mysterious properties, that provide a lot of insight that you often don't get from more boring systems. The hard ones make you think.

By spring 2011, Kepler had a sizable number of discoveries under its belt, but the number of planet candidates was growing much faster than the number of planet confirmations. There were so many systems needing examination that it would take decades to get through them all if we were to stick with the protocols of our early discoveries, using Doppler measurements or detailed dynamical studies to confirm their planet nature. In-depth dynamical studies can take several weeks or months to complete for a single system. Doppler measurements are even worse. Since most of the Kepler target stars were quite dim, it could take an hour or more per observation with the largest telescopes on the planet to make a single Doppler-shift measurement. Disentangling the Doppler signal for each planet from the ensemble of signals within the system—confirming their planetary nature and ruling out other astrophysical false positive signals—can easily require a few dozen measurements over the course of several years. This approach would not work for thousands of candidates. At the rate things were going, the mission would end with thousands of good planet candidates and only a few dozen confirmed planets.

For at least a year prior to launch, several members of the science team had been working on alternative methods of confirming planets that didn't rely on high-precision Doppler measurements. One approach involved systematically eliminating all of the non-planetary astrophysical signals. The most important type of false positive is some sort of eclipsing binary star blended into the same set of pixels as the target star. For any given planet candidate on a target star, there were a lot of possible blends that needed to be ruled out before it could be considered a real planet.

The variety of possible blends to consider is pretty astounding, and they have nicely technical-sounding names: hierarchical triple with stellar tertiary, hierarchical triple with planetary tertiary, background eclipsing binary with stellar or planetary tertiary, foreground eclipsing binary, etc. In general, they refer to distant eclipsing stellar systems that happen to line up with the target. Each of these scenarios comes with its own probability, which depends on the direction you are looking, as well as some of the properties of the target star. For example, interloping background stellar systems can cause different planet-mimicking signals for target stars with different brightnesses. Validating a planet candidate as a real planet with this approach requires eliminating all of these blend scenarios to a high degree of confidence.

Most of this work takes time on a computer and data from the spacecraft, rather than lots of Doppler measurements with a spectrograph. To gather the necessary supporting evidence, it takes only a handful of observations like a single, high-resolution spectrum of the star and high-resolution imaging. The high-resolution imaging can use either adaptive optics systems, where a deformable mirror removes the effects of the atmosphere, or *speckle imaging*, where you take a series of exposures that are so short, a few milliseconds each, that the effects of the atmosphere on the image are frozen in place and can be removed (the things that change from one image to the next are atmospheric, while the things that stay the same from one image to the next are astrophysical). These supporting observations place constraints on the presence of distant stars blended into the light from the primary target.

Other possibilities for blended scenarios can be eliminated with a careful analysis of the light coming from the Kepler spacecraft. The shape of the transit signal, for example, is slightly different for stellar eclipses than it is for planetary transits. Some false positives can be eliminated with a detailed look at the location of the center of the light from the target (the image *centroid*) when the candidate planet is in transit and out of transit. If the location of the centroid shifts in the sky during transit, it might indicate that the transiting object is not actually orbiting the target, but is instead orbiting an interloping object slightly to the side of the target.

Combining the high-resolution images, with data from Kepler, and models for the types of stars that exist in the galaxy, we can exclude blended scenarios and gain confidence that we're looking at a real planetary transit. This "Blender" analysis (named after the software package that was designed to examine these blend scenarios) would test all of the potential signals that could arise from the different possibilities against the observations—eliminating many millions of potential false positive signals. However, since it is impossible to eliminate all false positive signals, we needed to set a threshold where we had debunked enough of them to justify claiming a planet discovery. Where to draw the line is a somewhat arbitrary choice, but we went with 99.9 percent. If we could state with that level of confidence that a transit signal was not from a blended astrophysical source, we would *validate* the candidate as an actual planet. (On the Kepler mission, we distinguished between *validating* a planet and *confirming* a planet. To validate meant to eliminate other explanations, while to confirm meant that you had a constraint on the object's mass through some dynamical measurement—Doppler or transit timing variations. Either validation or confirmation would win your planet a spot on the discovery page, but the two approaches pointed to different sources of uncertainty.)

The first time we used Blender to validate a planet was in the Kepler-9 system. The Multibody Working Group published their study of Kepler-9 with two planet discoveries and one planet candidate. The two discoveries were the two planets near the 2:1 resonance whose gravitational perturbations were seen in the transit times. The planet candidate was the Earth-sized planet on a one-day orbit that turned up during the last minute search through the data—the final "gift" from this system. Rather than delaying the announcement of TTVs in this system, we left this final object as a planet candidate and turned it over to the developers of the Blender analysis, a pair of Harvard astronomers, Willie Torres and François Fressin. They were left to chew on that candidate and eventually validate it as a planet.

By now we were already two years into the mission, with only a dozen planets to show for it. Initially, Blender looked like it was the tool that we needed to rectify this problem. We expected that it could validate a lot of planets from the Kepler data, and fill up the planet list. But a full Blender analysis took a lot of computing time—twenty hours running on three thousand nodes of the Pleiades supercomputer at NASA Ames. That is sixty thousand hours of computer time—the equivalent of nearly seven years for a single processor (and that is after some streamlining improvements to the software).

Despite efforts to make Blender run more quickly, it became clear that it was a tool best suited to validating specific planets that were interesting, rather than a tool to study the thousands of planet candidates that remained unexamined. Blender did make key contributions to validating some iconic Kepler systems. The first example was the hot Earth, Kepler-9c. Another example was the smallest exoplanet discovered in the Kepler data, the Moon-sized Kepler-37b. Blender was used again to validate two planets in the five-planet Kepler-62 system. This included Kepler-62f, one of two planets from that system that were in the habitable zone of their host star. At the time it was announced in May 2013, Kepler-62f was the smallest planet (at forty percent larger than the Earth) that was discovered in the habitable zone of a distant star. These were all major contributions to the list of Kepler discoveries.

As Blender's throughput limitations became apparent, our Multibody Working Group looked at how TTVs might be used. A full-blown analysis of the perturbations to the orbital periods from TTVs also takes tens of thousands of hours. So just throwing the kitchen sink at a lot of systems wasn't going to speed things along any faster than Blender could. However, we didn't really need a comprehensive study just to show that the observed transits were due to planets. We only needed to be certain the objects producing the signal had masses in the planetary regime. It wouldn't even be necessary to show that they orbited the target star—after all, planets orbiting a background star were still planets. So we concocted a trimmed-down analysis that used some of the known properties of the TTV signal to cut out some unnecessary details, and that could be run in only a few hours. The plan we devised would first show that a pair of objects were orbiting the same star. Then we would show that the system would only remain stable if the masses of those objects were in the planetary regime.

We used the fact that when two planets are interacting with each other to produce a TTV signal, the changes in orbital period are generally anticorrelated. This means that when one planet's orbit is expanding (lengthening the time between planetary transits), the orbit of the other is shrinking (which decreases that time), and vice versa. Given a pair of planet candidates for a single target, we could use the anticorrelation of the changes in their orbital periods to show that they orbited in the same system by demonstrating that the anticorrelation was unlikely to be from noise.

With that step completed, we then used computer simulations of the system to place an upper limit on their masses. Limiting the mass of a planet is a much easier prospect than measuring it. To do this, we ran a suite of simple dynamical simulations of the planets and the central star. In the first simulation, we assigned the planet candidates gigantic masses which we knew would make the system unstable. Once it self-destructed, we started the simulation again, lowering their assigned masses. This process continued until we got below thirteen Jupiter masses for the planets (the traditional dividing mass between planets and brown dwarfs). If the system was still unstable when the objects had a mass that large, then we knew the objects had to have masses below that threshold and in the planetary regime, otherwise the system would have self-destructed a long time ago.

This two-step analysis was much faster than what we had done up to this point—using either Doppler measurements, a full TTV analysis, or Blender. A trio of us got to work on different approaches to implement this procedure: me, Eric Ford (from Penn State), and Dan Fabrycky (from the University of Chicago). What resulted was three papers that were drafted in late spring 2012, confirming the presence of twenty planets orbiting ten different stars. (I use "confirmed" instead of "validated" here because there were dynamical constraints on the masses of the planets.) These papers would constitute the largest haul of new planets that the Kepler Science Team had produced up to that point, including multiple planets in all of the systems from Kepler-23 through Kepler-32. However, before they could be announced to the public, they had to get through the mission bureaucracy.

These different methods to confirm or validate planets were new territory in exoplanet science. No one in the field had ever used them to claim planet detections before, and there remained concern about having something go wrong—necessitating an embarrassing retraction.

One safeguard that the Kepler team had in place was an internal review of all the scientific papers coming from the mission. Before a paper could be submitted for peer review, it would be read by a member of the Kepler Science Council. This council was a small cadre of science team members who oversaw the overall science program for the mission. It had a few permanent members, and a few additional members who were elected for terms of two years.

In late 2011 and through 2012, the volume of data coming from the spacecraft meant that many papers were published across the whole of the science team. Consequently, the science council members were exceptionally busy. Not only did they have their responsibilities to the mission to ensure that it continued to operate properly, but they had this responsibility to review all forthcoming scientific publications, and they had their own projects to work on. The result was long delays in getting the papers through the system and into the scientific literature.

The fact that many team members were trying new methods to confirm planets, when up to that time only Doppler measurements had been familiar (even a complete analysis of transit timing variations was suspect), meant there was a lot of trepidation with just publishing quick analyses of large groups of systems with only upper limits on the planet masses. The problem we faced with introducing these new methods to confirm planets was that those of us who applied them were also the ones with the preexisting expertise about the underlying physical processes that justified the planet confirmation claims. When questions arose about the validity of some aspect of the study, it often took more than the available few minutes in a given meeting. Terse answers to questions from name-brand scientists isn't the way to instill trust. Going into depth about technicalities that would otherwise consume a day or two in a planetary dynamics course wasn't an option either. There just wasn't time.

For several months in late 2011 these new results were stuck in a log-jam. The Kepler Science Council were busy, and were being asked to approve a set of papers that would use an unfamiliar method to double the number of Kepler planet discoveries. Having some ballast when urgency strikes is not a bad thing since one can cause a lot of destruction if rushed into a decision under duress. However, it was clear that something needed to change if the mission were to fulfill its potential in producing results.

The issue came to a head on a phone call with a large fraction of the science team and the science council. After our group repeated our methodologies and results in these papers, and reiterated our confidence in them, we laid out our case regarding the limitations of the current organizational structure for the mission, specifically regarding the internal reviews of new results. The members of the science council had too many responsibilities and it was nearly impossible for them to accomplish all that was expected of them. They couldn't keep up with the publications that were coming from the working groups. If Kepler wanted to make a lot of new discoveries, it needed to streamline the publication process.

The science council eventually divested itself from the responsibility of internally reviewing each manuscript—pushing it down to the working groups, who were now charged with internally reviewing the papers that originated from them. This situation, and its resolution, had a profound effect on me personally, and shaped my views on how to manage a large collaboration. It showed me the downside of concentrating too much responsibility into so few people. It simply was not possible, or fair, to expect the members of the science council to accomplish all the tasks that were heaped upon them. If we wanted to function going forward, we had to trust the smaller organizations within the mission, the working groups, each other, our skills, and our integrity as scientists.

These three papers, each with a slightly different approach to using TTVs and dynamical stability to confirm planets, appeared in 2012. They were a significant boost to the growing list of planet discoveries. Following on the heels of these papers was another pair of studies, one by me and another by a scientist outside the Kepler team (Jiwei Xi, currently at Nanjing University), that used the same technique to confirm an additional forty-one planets in twenty different systems. These later two studies were done independently. We only found out about each other when they both appeared a day apart on the arXiv preprint server—where drafts of scientific papers are collected for the broader community to see. I had analyzed thirteen systems and Jiwei analyzed twelve, with five systems common between us. (Jiwei and I have similar interests, and this would not be the last time that we pursued similar studies at roughly the same time.)

While this new approach to planet confirmation had nearly quadrupled the number of planets for the mission, the candidate list was still outpacing us. However, accompanying these papers that used TTVs to confirm pairs of planets, our group submitted a separate study that had the potential to make up ground on the burgeoning candidate list. The approach we devised also took advantage of the multitransiting systems. The basic idea is that, while a single planet might have a modest probability of being a false positive signal due to eclipsing binary stars or noise, the probability of having two false positive signals on the same target is much smaller. And the probability of having three false positives is minuscule. So systems that show multiple transiting planet candidates are almost certainly real planets because the chances of having multiple false positives on the same target are so small.

The first time we applied this *validation by multiplicity* method was with the Kepler-33 system in an analysis led by Jack Lissauer. Here, five planets in a single system were validated based on how unlikely they were to be false positives. Once the method was shown to work, it took some time to shore up the mathematical formalism and to select from among the Kepler planet candidates the systems that were suitable for this kind of analysis. After about a year and a half, our group wrote up a pair of studies, led by Jack Lissauer and Jason Rowe, that used this approach to validate 851 planets in 340 systems—a huge increase in the number of bona fide planets in the Kepler data, and a number that was almost twice the number of previously known planets across the entire field of exoplanets. That was 2014. Two years later, after the Kepler mission had ended, Tim Morton at Princeton developed a software package that was like a streamlined version of Blender (this one called VESPA, which stands for Validation of Exoplanet Signals using a Probabilistic Algorithm). He used that software to validate some twelve hundred additional planets, again nearly doubling the number of previously known planets.

The multiplanet systems seen by Kepler, even including those that remained planet candidates, enabled investigations into the correlations among planets in the systems, such as determining trends in planet sizes or orbital period ratios. We've seen that the effects of resonances in a planetary system can be pronounced, such as the mean-motion resonance between Neptune and Pluto, or the overlap of the two secular resonances of Mercury and Jupiter. There are even resonances that show up in the moon systems of different planets. For example, three of the four large moons of Jupiter are in a three-body, mean-motion resonance. Io, Europa, and Ganymede orbit Jupiter with periods of 42.5, 85, and 170 hours—giving ratios of 1:2:4.

We know that this resonance structure likely originates from object pairs forming within a gas disk. For Jupiter, the gas disk caused the moons to migrate relative to each other, allowing them to capture into this state. If a set of bodies does capture into resonance, they tend to remain stable—even in the presence of some external disturbances. However, if a set of objects gets close, but doesn't actually capture into the resonance, it makes the system more likely to go unstable. This, too, shows up in the solar system. When we look at the rings of Saturn, we see large gaps between the different rings. Those gaps were first identified by the Italian-turned-French astronomer, Giovanni Cassini, in 1675. The cause of this gap was found a century later, in 1789, by William Herschel—namely, the small moon Mimas.

If you were to place a small particle in the largest of the gaps in Saturn's rings, the Cassini division, it would orbit roughly twice for every one orbit of Mimas. If you place it just right, it would be trapped in the 2:1 mean-motion resonance with Mimas and remain stable. Indeed, the Cassini division is not completely devoid of ring particles—many exist in the gap, with orbits in stable resonance with Mimas. However, most of the orbits in the vicinity of the resonance are unstable. So any ring particle that misses a resonance will eventually be removed from the area, as it scatters out of the neighborhood through dynamical instability. It is this depletion of unstable ring particles that produced the gap. Several of the other gaps in Saturn's rings are caused by these same kinds of interactions.

A similar effect occurs in the asteroid belt. The main asteroid belt is a ring of debris between the orbits of Mars and Jupiter. The region spanned by the asteroid belt includes several mean-motion resonances with Jupiter. Objects not trapped in the resonance generally have unstable orbits, and the region becomes depleted in material. The result is gaps in the main belt asteroids called Kirkwood Gaps—discovered in 1866 by the American astronomer, Daniel Kirkwood.

Some groups of asteroids do survive because they happen to live in a mean-motion resonance with Jupiter. The Hilda asteroids, for example, are trapped in 3:2 resonance with Jupiter—each Hilda asteroid orbits 3 times for every 2 orbits of Jupiter. While the individual orbit of a Hilda asteroid is an ellipse, the conjunctions between a Hilda asteroid and Jupiter, when they line up with each other and the Sun, all occur in the same part of the orbit of the Hilda. They always occur when a Hilda is near its perihelion—when it is closest to the Sun, and therefore the farthest it can be from Jupiter while still lining up with it and the Sun. Because these are the only stable orientations of their orbits, the collective aphelia (where they are farthest from the Sun and moving most slowly) of the individual elliptical orbits form an equilateral triangle that constantly points away from Jupiter. In the past, there were likely many more asteroids in this part of the solar system, but those that didn't live in the few stable orbits were forcibly removed from the system.

Perhaps the most famous group of trapped asteroids in the solar system is the Trojan asteroids. These are contained in 1:1 resonance with Jupiter, meaning that they orbit along with Jupiter—at the same distance. Trojan orbits exist because the gravitational influences of Jupiter and of the Sun, combined with the motion of Jupiter around the Sun, carve out a place where all three influences cancel—leaving a stable equilibrium point where an object drifts along with Jupiter, staying constantly in that part of the sky. There are actually five such points for every orbiting object, called Lagrange points after one of the mathematicians who discovered them (Joseph-Louis Lagrange and Leonhard Euler).

The Lagrange points L1, L2, and L3 are unstable, meaning that if an object is located there, a small displacement from the correct location will start driving it away from the Lagrange point. You can find them along the line between a planet and the central star. L1 is between the planet and star where its faster orbit is slowed just right by the gravitational tugging from the planet, L2 is just exterior to the planet's location where its slightly slower orbit is sped up by gravitational tugging, and L3 is on the planetary orbit, but opposite the planet's position, like a mirror image of its position relative to the star. The remaining two Lagrange points, L4 and L5, are stable, where small displacements from the equilibrium will drive the object back toward the starting point. These two points are located on the planet's orbit, but 60 degrees ahead and 60 degrees behind, forming equilateral triangles with the star and planet. Figure 5.1 shows the location of the Lagrange points for a planet orbiting a star.

In recent years, the L2 Lagrange point for the Earth has risen in fame because it is where NASA's James Webb Space Telescope orbits the Sun, staying fixed relative to the Earth.[2] Since the L2 point is unstable, it requires constant, small adjustments to prevent the telescope from drifting too far from the nominal location. The Trojan asteroids swarm around Jupiter's two stable points L4 and L5. Estimates suggest that there could be a million asteroids trapped in the Trojan orbits of Jupiter. A handful of similar objects have been found for Venus, Earth, Mars, Uranus, and Neptune.

There has always been some interest in searching for Trojans in exoplanet systems. An early paper on the subject came from Eric Ford, from Penn State, and Scott Gaudi, from The Ohio State University, which outlined one method to find them. Here, you compare the orbit for a

---

2. The James Webb Space Telescope is a large infrared space telescope that was launched on Christmas Day, 2021. While it is often touted as a successor to the Hubble Space Telescope, a more appropriate comparison is as a successor to the Spitzer Space Telescope—the infrared counterpart to the Hubble and one of the four *Great Observatories* (Compton for gamma rays, Chandra for X-rays, Hubble for visible light, and Spitzer for infrared).

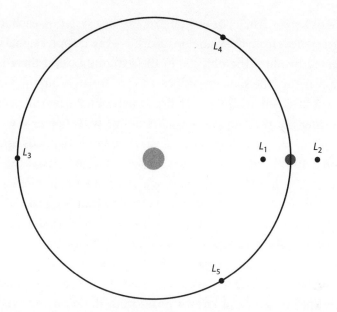

FIGURE 5.1. The five Lagrange points are locations in a planetary orbit where the gravitational influence of the planet, the central star, and the orbital motion of the planet effectively cancel and the planet and the object will stay in fixed relative positions as they orbit. The three collinear points (L1, L2, and L3) are unstable. Small disturbances to the orbit of an object located there will cause it to drift away from those locations. The two equilateral points (L4 and L5) are stable, and small disturbances will not cause an object located there to drift away. Instead, it will circulate around the Lagrange point as it orbits the central body.

planet that you measure using the Doppler shift to the orbit that you measure with transits. If the planet has a Trojan, the transit signal will not line up with the Doppler signal. The Doppler signal depends upon the distribution of all of the mass orbiting at a given orbital period. It arises because the star is offset from and will orbit the center of mass of the entire system. The presence of a single planet displaces the star from the center of mass, and the star's orbit about that center yields the Doppler signal. If there were a ring of planets orbiting at the same distance, the star would no longer be displaced from the center (because there is no longer an imbalance in the mass distribution from the planets), and the Doppler signal would disappear.

If there were a Trojan, or several Trojans, in a system—each located 60 degrees away from the planet in its orbit—the Doppler signal would be affected by all of the objects. In the extreme case, if there were a second planet of the same mass located in a Trojan orbit, the Doppler signal would line up with a point halfway between the two planets. This displacement in the Doppler signal is unlike what happens with the transit signal for the planet. The transit signal only depends upon the position of the planet relative to the host star. When it passes the line of sight, it blocks the star regardless of whatever else might be in the system. So, if you see a Doppler signal offset from a transit signal, it tells you that there is an imbalance in the mass somewhere along the planetary orbit that is shifting the Doppler signal relative to the transit signal.

Another way to find Trojans is to simply look for them to transit the host star directly. They would do so with the same frequency as a known planet, and at a known position in the orbit relative to the known planet—one-sixth of the orbit ahead or one-sixth behind. The challenge here is that Trojans tend to be much smaller than the orbiting planet, so the signal would be very weak. So far, no searches using either of these methods have turned up positive results.

While there have not been any exoTrojans detected around a distant star, there was an exciting false alarm. Early in the mission it appeared that the KOI-730 system had two planets with the same orbital period circling a star similar to the Sun. Co-orbital Trojan planets like this would be an unusual situation. But, as we've seen with WASP-47, the hot Jupiter with neighboring small planets, unusual situations can give a lot of interesting insights into planet formation and dynamical evolution. Had this pair of planets been real, it would have forced planet-formation theorists to explain how it got there and why that process, whatever it may have been, produced so few examples.

For a few months, the co-orbital planet model for KOI-730 made the rounds among scientists as the data were examined and ideas tossed around as to how the system could have formed. Word got to the press, and a few articles surfaced about the possibility of a pair of co-orbital planets. However, after some intense scrutiny of the Kepler data, we found that the orbital period of one of the planets had been measured incorrectly. Instead of two planets with the same period, one of the orbits was twice as long as originally thought. Perhaps it was some noise in the data that looked enough like a transit for the computer to think that the shorter period was correct. But, as data continued to arrive and the system got more human attention, the true orbital period persisted while the spurious one faded back into the noise.

As cool as it would have been to find a pair of Trojan planets, the actual properties of the system were interesting in their own right. The corrected orbital period for the misdiagnosed planet meant that the four planets in KOI-730 (now renamed Kepler-223) had period ratios of 8:6:4:3. This set of ratios of orbital periods contains a lot of resonance behavior. The inner pair of planets and the outer pair of planets both have period ratios of 4:3. The first and third planets, and the second and fourth, have period ratios of 2:1. And the middle pair has a period ratio of 3:2 [48]. This complex network of resonances didn't come about randomly. It was almost certainly the consequence of their mutual migration within the system. As the planetary orbits converged toward one another, they captured into this configuration until the disk of material dissipated—leaving them trapped where they were. Kepler-223 is not unique with its collection of resonances, but it is rare. Only a small fraction of planetary systems have planets that are actually in a resonance, as opposed to simply being near one. And a small fraction of that small fraction has multiple resonances across the several planets. Two other examples are the Kepler-80 system with six planets, and TRAPPIST-1 with seven.

For Kepler-80 (originally KOI-500), the Kepler team initially found five planets, all with orbital periods less than ten days. Five years later, a pair of scientists (one working at Google) used a machine-learning algorithm to find a sixth planet buried in the data—this one with a two-week orbital period. The inner planet in Kepler-80 is somewhat detached from the other five (it may not come as a surprise that it is another one of those Earth-sized planets on a one-day orbit), but the outer five planets in Kepler-80 are in a chain of resonances, where all of their orbits are strongly coupled to each other—having period ratios of 3:2, 3:2, 4:3, and 3:2 [49].

The planets in TRAPPIST-1, which also has a complex resonance structure, have period ratios of 8:5, 5:3, 3:2, 3:2, 4:3, and 3:2 [50, 51]. Nowhere in the solar system do we see systems of planets or moons whose orbits are this intertwined. The nearest example is the 4:2:1 resonance structure that we see with three of the four large moons of Jupiter (Io, Europa, and Ganymede), or perhaps the 33:22:18 ratios seen in the small moons of Pluto. But these exoplanetary systems participate in much stronger interactions due to the fact that their resonances are so close together, such as first-order resonances like 2:1, 3:2, or 4:3.

These interesting systems notwithstanding, the majority of Kepler planets aren't near any strong resonances. But the distribution of orbital period ratios from Kepler is neither uniform nor random—there are some orbital relationships that appear more common than others, and some that are strikingly less common. When we looked at the distribution of orbital periods, one of the first features we noticed was an overabundance of planet pairs that are wide of the first-order, mean-motion resonances. The peak in the abundances is within about five percent of those values, with the most prominent ratio being the planet pairs slightly wider than 3:2 (the same ratio as Neptune and Pluto). At the same time, there are almost no planet pairs that are just interior to those same ratios.

It isn't clear why these excess planet pairs are there, and why they are all wide of the resonance instead of being short of it. Most theories point to the idea of migration in the disk as the planets are forming, which can either drive the planets near the resonances but not capture them, or drive them into the resonance and then the resonance is later broken—either because of interactions with debris left over from the formation, or from other planets in the system. What doesn't happen is planets being driven just past the resonances and staying there, or planets spreading apart and stopping shy of the resonances. Resolving the question of what causes so many planet pairs to be just wide of resonance remains the subject of ongoing research.

When comparing the Kepler planets to the planets in the solar system, we see elements of shared experiences, along with many striking differences. We know from the period ratios we see in many exoplanet systems, and in the solar system, that planet migration is common in one form or another. Mean-motion resonances can result from this migration, but they are less common—especially at planetary scales. A lot of mean-motion resonances in the solar system occur among the smallest objects, like moons, rings, and asteroids, which Kepler would not have been able to see. The most striking differences between Kepler systems and the solar system are the sizes of the planets and the sizes of the orbits. Most Kepler planets are between the size of the Earth and the size of Neptune. There are no examples in the solar system of planets in this range. The overwhelming majority of the Kepler planets orbit interior to the orbit of the Earth, and many would be well interior to the orbit of Mercury.

There remain a number of unanswered questions resulting from these discoveries. In particular, why didn't the solar system turn out to be like the Kepler systems? Given how common these planetary systems appear to be, some have speculated that perhaps all planetary systems started like those seen in Kepler and the unusual systems are those that somehow get rid of their first set of planets. Perhaps the initial set of planets is driven into the central star as the larger bodies in the system migrate around before settling into their final orbits. Perhaps the initial set is destroyed as the system becomes unstable as the initially closely spaced planets interact with each other gravitationally. Then, after self-destructing, the system would either start over with a second batch of planets, or have some remnants that survive—which would be more like the terrestrial planets that we have orbiting the Sun.

Either way, there must have been some difference in our past that prevented Earth and our sibling planets from congregating deep in the inner solar system. There must have been some difference that either prevented the inner planets from growing into the super-Earths and sub-Neptunes that seem to be common in other systems, or removed them from the system had they originally been here. As we continue to study the systems from Kepler and other exoplanet missions, we may hopefully gain a clearer understanding of the paths that the solar system didn't take and how worn those paths are.

6

Strange Stars and Star Systems

A lot of effort on the part of the science team was devoted to eliminating false positive signals caused by binary star systems and other astrophysical, transit-mimicking sources. A similar effort on the part of the science team was to ignore the exoplanets entirely and to treat those meddling binary stars and astrophysical sources as the signal to study. Stars have spots, they shake and pulsate, they often come in multistellar systems that eclipse each other. All of these effects can show up as small brightness variations of the system, which is precisely what the Kepler satellite was designed to measure.

With the twenty-parts-per-million photometric precision of the instrument, there was bound to be a lot of strange stuff that appeared, and indeed, there was. One of the first strange systems was introduced to the science team at the 2010 Denmark meeting. It was a star that, at regular intervals, would quickly brighten, and then dim again—almost like a reverse transit. It was so strange that we weren't even sure how to categorize it—we weren't sure whether or not it belonged on the KOI list. It was clearly an interesting signal, but it was clearly not a planet. Eventually, Ron Gilliland (one of the stellar astrophysicists on the science team) pointed out that we all found it interesting, so it certainly qualified as a *Kepler Object of Interest* in the strictest sense of the phrase. So it was added to the list.

KOI-54, as it was called, would brighten by about seven percent every forty-two days. The signal was so strong that there was no way to brush it aside as noise or some other anomaly. The light curve was a beautifully repeating pattern of prominent peaks. One of the first hypotheses we considered for the origin of this signal was that it might be a black hole orbiting the target. The idea was that when the black hole transited the star, the strong gravitational field of the black hole would bend the starlight and focus it onto the telescope—making it appear brighter. Alas, the black hole idea didn't pan out, but it shows the lengths we were willing to go to explain what we saw.

I should mention that this idea of a compact object gravitationally lensing a companion star was not without merit. A few years after this meeting, Eric Agol flexed again, this time with his graduate student Ethan Cruse. They found a system which appeared to show a planetary transit, with the target star getting dimmer. But halfway between transits the star would brighten by a similar amount. That system, KOI-3278, turned out to be a binary star system that contained a Sunlike star and a *white dwarf*. (A white dwarf is the leftover corpse of a star that has roughly the mass of the Sun squeezed by gravity into a sphere the size of the Earth.) The dimming *transit* in this system turned out to occur when the white dwarf star went behind the target star and its light was blocked (not a transit at all, but rather the eclipsing of the small white dwarf) [52]. The brightening signal in this system occurred when the white dwarf passed in front of the larger star and the white dwarf's strong gravitational field bent the light from that star, focusing it onto the telescope, and making the target appear brighter.

Back to KOI-54. A closer examination of the system showed that, not only did the system have these periodic seven-percent brightening events, but the brightness also bounced up and down by roughly one percent in a repeating pattern between each of the larger events. The team took additional data with ground-based Doppler spectrographs as a means to rule out some of the models that were proposed to explain the system. Eventually, we learned the truth about this system

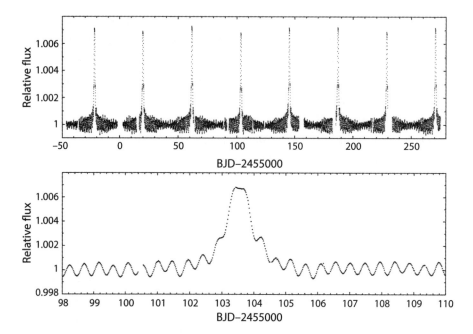

FIGURE 6.1. Brightness variations in the KOI-54 system as seen by Kepler. Two large stars orbit each other on highly eccentric orbits. When they are close together, the distortion to their shape and their mutual heating cause the brightness of the pair to spike. Between encounters, the stars pulsate, which gives the smaller oscillations in the system brightness. *Source*: AAS Journals.

in a paper led by Bill Welsh and Jerry Orosz, the pair of astronomers from San Diego State University who studied the ellipsoidal variations and Doppler beaming in the HAT-P-7 system. It turns out that KOI-54 is a pair of high-mass stars orbiting each other with highly eccentric orbits (eighty-three percent) [53]. Figure 6.1 shows the brightness of this system seen in the Kepler data.

As the pair of stars made their close approach, the intense radiation from each star would heat the surface of the other. The higher temperature of the opposing faces of the stars increased their combined brightness by roughly half of what was observed. The other half of the brightening occurred because, as the stars made their close approach, the tidal forces would distort the star, causing ellipsoidal variations— the distortion of the star into the rugby-ball shape that we also saw from HAT-P-7. In this case, however, the distortions weren't caused by a planet that was a thousand times smaller than the Sun, but by another star that is nearly two and a half times more massive than the Sun. The large masses of the stars caused a large distortion in their shapes. Because a larger portion of the star's surfaces were now visible, they appeared brighter—contributing to the other effects brightening the system.

After the stars pass through their close approach and begin to recede from each other, the distorting tidal force quickly weakens. As they drift apart, the two stars attempt to settle back into a spherical shape, but they overshoot their mark. The result is that the stellar material keeps sloshing around—pulsating back and forth between the shapes of a rugby ball and an M&M (prolate and oblate spheroids). It is as though you have a pair of water balloons that bounce off each other and start vibrating in response. While the stars in this system don't actually contact each other, the gravitational force from the close approach causes a similar change to their shape. Each of the two stars gets smooshed and stretched, back and forth between fat and skinny ellipsoids. They do this ninety and ninety-one times between each successive orbit—lining up perfectly with the next close approach. The system has settled into a strong coupling between the star-sloshing frequencies and the orbital frequency.

KOI-54 was not, apparently, the first system that we knew about with this kind of behavior. It was the fourth. But it was by far the best characterized of the four because of the quality of the Kepler data. It also became the first of several so-called heartbeat stars that Kepler found. The light emanating from these heartbeat stars resembles an electrocardiogram with strong pulses separated by relatively quiescent intervals—hence their name. There are now a few dozen heartbeat stars in the Kepler data that have been studied, none quite as pristine as the KOI-54 system (in my opinion), but still interesting in their own right. All of these systems arise from the brightness changes caused by a pair of stars orbiting each other on eccentric orbits, heating each other's surfaces, distorting each other's shapes, and oscillating like gelatinous blobs as the distorting tidal forces subside.

The relatively small pulsations of heartbeat stars, with brightness changes of one to a few percent, show up prominently in the Kepler data. But they are not the only sources of pulsating changes to stellar brightnesses—nor are they the most common type of variable star. Several types of waves propagate throughout stellar interiors, which also cause changes in stellar brightnesses. Studying the properties of these waves, measuring their oscillation frequencies (or wavelengths), and the relationships that they have with each other, provides a means to characterize stellar interiors. They give us a window to look at the goings-on deep below the surface of the star. This is incredible when you consider that you are looking at waves with wavelengths that are several times larger than the Earth, bouncing around the inside of a distant ball of gas.

This study of asteroseismology is similar to how we use different waves produced from earthquakes to study the properties (density, mass, pressure, etc.) of the interior of the Earth. We measure the wavelengths and frequencies of the seismic waves moving around the Earth's interior, the speed with which they move, and reflections that occur at the boundaries between different layers in the Earth (say, the inner core and outer core, or the core, mantle, and crust). Piecing together the information from several types of waves that originate across the Earth's surface is one of the few ways that we can probe the Earth's interior.

Studying the shaking of the Sun, helioseismology, can tell us similar information for the interior of the Sun—whether it has layers, how thick those layers are, and the properties of the gas in those layers. There are a variety of waves that travel throughout the Sun. The two most important for our purposes are p-waves and g-waves, or p-modes and g-modes, or *acoustic* modes and *buoyancy* modes. (These names are given for reasons that we will shortly see.) P-modes are pressure waves, which propagate through the Sun because of pressure fluctuations in the gas. They are essentially the same as sound waves in the Earth's atmosphere. P-modes can move throughout the entire interior of the Sun, from the core to the surface.

Gravity waves, or g-modes, are more like water waves than sound waves. Here, gravity is the force that drives the oscillations. When we look at water waves, if there is a bulge of water poking up above sea level, gravity pulls it back down. Eventually, it overshoots and causes a dip in the surface, while the surrounding water gets pushed upward—forming bulges that gravity will subsequently pull back down. The process repeats itself and produces the waves that we see. G-modes in the Sun propagate deep in its interior. Measuring the properties of both p-modes and g-modes of oscillation tells us what is going on at these different depths beneath the surface.

While the Sun is a noisy place that shakes at a wide range of frequencies, the thicknesses, temperatures, and pressures in its different layers make it so that the Sun resonates with the noise at some frequencies, while noise at other frequencies does not resonate. In other words, the Sun has certain frequencies where it naturally likes to oscillate, and others where it does not like to oscillate. Random noise that exists at all frequencies will be amplified for those frequencies that resonate, and will dissipate for the frequencies that don't resonate. Brightness variations, which is what Kepler is looking for, will be more prominent for resonant frequencies than for non-resonant ones.

If you look at how the Sun's brightness changes at different frequencies, you find a discrete set of them where the Sun's brightness changes most dramatically. The most prominent peaks in the helioseismology spectrum are the p-modes, which oscillate at several, equally spaced frequencies surrounding a value of 3 millihertz. This corresponds to fluctuations every few minutes as sound waves compress and rarify the hot gases in the solar atmosphere. In compressed regions the temperature rises slightly and the brightness climbs. Rarified regions have lower temperatures and are dimmer. The associated variations show up in the brightness over time.

Kepler observations of distant stars also detect these oscillations. Indeed, some of the Kepler data for distant stars rival the quality of data we get for the Sun. By studying the asteroseismology signal from Kepler stars, we can determine their interior properties just as we have done for the Sun. Stellar astrophysics is an important subject in its own right, so the fact that we can use Kepler data to learn about stars is a bonus. Moreover, understanding the properties of stars is a key ingredient to understanding the planets that orbit those stars since, in many cases, we are limited in our understanding of the properties of planets by our knowledge of the properties of the host star, rather than measurement errors from transit or Doppler data.

Transit measurements only give us the relative sizes of the planet and star, not the absolute sizes of either. Doppler measurements only give the ratio of the masses of the planet and star, not the absolute masses. To find the actual size and actual mass of the orbiting planets, we need separate information to determine the stellar size, mass, temperature, and age. Some of this work can be done by an analysis of the stellar spectrum (details about the prominence of certain spectral lines and the shape of those features contain some of that information). In other cases, we must resort to putting what observational information we can into a computer model and just using the results. With Kepler, good asteroseismology targets provide the highest-precision estimates for these stellar properties. Improved measurements of stellar properties, in turn, improve our estimates for planetary properties.

Some of the things Kepler can tell us about stars are quite astonishing. For example, the rotation of a star causes a Doppler shift in the frequency of the stellar oscillations. The portions of the star that approach us will manifest a higher oscillation frequency, while those portions that recede from us will have a lower frequency. The result is that the observed oscillations will *split* into nearby frequencies that differ by that stellar rotation frequency from where the star would naturally oscillate if it were not rotating.

This rotation is only visible if there is a portion of the star that is actually approaching or receding from you. If you were to observe the pole of the star, where all of the motion is perpendicular to the line of sight, then the rotational splitting would vanish. Since there are other ways to measure the rotation rate of the star aside from asteroseismology, the mismatch between the rotation rate and the observed oscillation splitting can tell you which latitude of the star is beneath your gaze. That is, you can tell whether you are looking at the stellar pole, or its equator, or anywhere in between.

The ability to observe stellar rotation in this manner has other ramifications. Because the g- and p-modes probe different depths in the stellar interiors, we can separately look at the rotational splittings of the g- and p-modes in the asteroseismology signature to see whether the inner parts of a star rotate at the same rate as the outer parts of the star. One might think that all parts of a star would rotate at roughly the same rate, but that isn't what happens. Since a star is a ball of gas, it flows like a fluid rather than spins like a top. Some parts flow at different rates from others. The Sun's equator, for example, flows at a different rate from its poles, with the equator completing a rotation in twenty-five days while the poles complete one in thirty-five days. This *differential rotation* is a common feature in rotating fluids.

As you move toward the interior of a star, the different layers may not be strongly coupled to each other—meaning that there may not be much friction between them. Without a strong connection, a fast-rotating inner layer will not be able to drag a slower-rotating, outer layer into lock step. The result is different rotation rates at different depths within the star. We see this on the Earth as well. The atmosphere is not strongly coupled to the surface, and so it can move relative to the surface, producing the jet streams. As a star evolves, the core of the star will shrink and spin faster (like an ice skater pulling in their arms to spin faster), while the outer layers expand and spin more slowly. This process can further decouple the rotation of the inner parts of a star from its outer portions.

With Kepler data we can see this effect develop in stars. During the main-sequence portion of their lives, stars like the Sun have reasonably strong couplings between their inner and outer layers, and the different parts rotate at similar, though not necessarily identical, rates. The Kepler data are sufficient to show the differential rotation in Sun-like stars, as the equators rotate at slightly different rates from the poles. But it also shows that in aging stars, the inner layers rotate at different rates from outer layers—the g-modes of the deep interior have a larger rotational splitting, implying a faster rotation, than the p-modes near the surface. This is what you would expect if the inner part were shrinking in size and the outer part expanding.

Sound and gravity waves aren't the only kinds of pulsations that stars can undergo. In fact, there is a large population of stars that pulsate wildly. These variable stars live in a long and narrow range of related stellar temperatures and brightnesses. (This range is called the *instability strip*.) For some of the stars in this range, the temperature in the stellar interior becomes hot enough to ionize a layer of gas. That layer becomes more opaque to light that is trying to escape the core, just like how a bolt of lightning (itself a cylinder of ionized gas) is opaque and doesn't allow us to see the sky behind it. The ionized gas also stores energy in the newly liberated electrons.

Now that the light and energy are trapped in the stellar interior, they exert a pressure on the opaque layer, pushing it away from the core. The expanding layer drives outward the material that is above it. As the star expands, its outer layers cool and become more diffuse. Eventually, the temperature drops to the point that the delinquent layer is no longer ionized, meaning that it is now more transparent. This transparency allows the light to escape the star. It also releases a bunch of energy, as the formerly free electrons recombine with the ions in the gas. This release of energy causes the star to brighten.

Since there is no longer the additional pressure on the formerly ionized layer of gas, the outer layers of the star, including the aforementioned layer, collapse again toward the core. As the layer collapses, it heats up, which eventually causes it to ionize again. This cycle repeats itself at a consistent rate, beginning with an opaque layer of ionized gas trapping energy; through the process where that light escapes by driving the ionized layer out toward the stellar surface where it cools, becomes more transparent, and releases the energy stored in the free electrons as they rejoin the atomic nuclei; to the point where the layer contracts, heats up, ionizes, and becomes opaque again. For some stars, especially the smallest ones, the cycle occurs in minutes or hours, while in other, larger stars the cycle can take months to complete.

These variable stars (they come in a few flavors) have played key roles in our understanding of the universe as a whole. The period of the pulsations correlates with the intrinsic brightness of the stars—a relationship first identified in 1908 and 1912 by Henrietta Leavitt, one of the several pioneering women astronomers who worked as "computers" at the Harvard Observatory. She found the relationship between the pulsation period and brightness in the stars of the Small Magellanic Cloud—a small galaxy that orbits the Milky Way galaxy like a natural satellite (though this wasn't known at the time). Once her relationship was calibrated, astronomers could then compare the apparent brightness of the stars to their intrinsic brightness from Leavitt's relationship. With that information, they can measure the distances to them because, given two identical pulsators, the one that appears brighter will be closer to us than the one that appears dimmer.

About a century ago, Cepheid-type variable stars were seen by the astronomer Edwin Hubble inside the Andromeda galaxy—the nearby sister galaxy to our own Milky Way. Because of the relationship that Leavitt identified between the brightness of these stars and their pulsation periods, Hubble was able to measure the distance to Andromeda. From these observations, he showed that Andromeda was not part of our own galaxy, like some planet-forming spiral nebula, but was rather a galaxy in its own right and much more distant than astronomers had previously imagined.

A challenge with studying variable stars is that their pulsations occur on a variety of timescales, some of which can correspond to the natural timescales of the Earth. It doesn't make much difference for a ground-based observer whether the entire cycle of brightening and dimming takes place over a few weeks or months or years, and nearly doubles the brightness of the host star in a given cycle. This is the case for the largest and floppiest of the stars. For the smallest and densest pulsating stars (white dwarf stars), the brightness variations may only be one percent. Fortunately, for these small stars, the pulsations can occur in a few minutes and can be observed multiple times in a single night.

The worst-case scenario is a cycle that happens over twenty-four hours, and produces a small variation. It can be a challenge to separate the signal that you are seeing from the rotation of the Earth. These measurements can be further complicated when the rotation period of the star, not just its pulsation period, happens to nearly match the rotation of the Earth as well. Kepler's continuous observing overcame the confusion that arises because of the Earth's motion.

Kepler's observations of variable stars produced unprecedented insights into their interiors, and into the physics that drove their behavior. Consider that in the decades prior to Kepler, a network of ground-based telescopes, called the Whole Earth Telescope, were set up to study pulsating white dwarfs. Despite the constraints of ground-based observing, they made significant progress in monitoring these stars— gathering state-of-the-art data on white dwarfs at a rate of two targets, observed twice per year, for two weeks at a time. Kepler, by comparison, was able to observe about forty such objects, continuously, for years. In its four-year mission, Kepler obtained the equivalent of two and a half centuries of ground-based data.

Stellar astrophysics with Kepler was no joke.

With the asteroseismology data, Kepler measured a wide array of properties for a wide variety of stars. It was able to deduce the separate layers of oxygen, carbon, and helium inside white dwarf stars, because the strong gravity of these compact objects causes them to differentiate by the elemental masses. Kepler was able to distinguish between different kinds of cores in giant stars—specifically, ones where the nuclear fusion of helium was ongoing in the core, and ones where fusion in the core had ceased and only occurred in a shell of material surrounding the core. Kepler could measure the tilt of the stellar rotation axis of stars relative to the line of sight—thereby constraining one of the angles between the rotation axis of the star and the orbital axis of its planets [54]. One such planetary system was found to have orbits tilted nearly 45 degrees relative to the star's spin. (It takes two angles to uniquely specify the star's spin axis in space and asteroseismology gives one of the two angles—the tilt of the stellar rotation axis from the line of sight. The other angle, as we've seen, is measured with the Rossiter–McLaughlin effect, looking at the anomalous Doppler shift that occurs while a planet transits the star.)

Another area of stellar astrophysics where Kepler made major strides was measuring the ages of different stars. Stellar age estimates are notoriously difficult. In some cases, the measurement uncertainties are almost the same size as the age itself, like "five billion years, give or take four billion." One can make reasonable estimates for the age of some stars using asteroseismology because the density of the core changes as the hydrogen is fused into helium, and the density of the outer layers evolves in response to the changing core. These changes affect the frequencies of the p-modes and g-modes. However, not every star is conducive to that analysis, especially if the stars are too dim to tease out the seismic signal. In many cases (especially for stars not observed by Kepler), it is difficult to collect sufficient data, of sufficient quality, to do what Kepler can do when monitoring its best and brightest targets.

One participating scientist on the Kepler team, Søren Meibom, had a program to use Kepler data to calibrate an alternative method to estimate stellar ages. Here, Søren studied the rotation properties of stars to see how a star's rotation is connected to its age. He was looking specifically at stars that were members of large star clusters where the ages of the clusters (and hence the stars they comprised) were already known. Star clusters like these have always been an important resource in the history of stellar astrophysics because all of the stars in the cluster form at approximately the same time, and they all have approximately the same initial chemical composition. For *field stars*, which are randomly distributed in the sky and don't share a common background with their neighbors, the different chemical compositions and different stellar ages can make it difficult or impossible to extract many details about their histories. The similar age and composition of star cluster members eliminates many of the obstacles that prevent the measurement of these details.

Measuring stellar ages through rotation, or *gyrochronology*, relies on some universal physical effects that occur for stars that are roughly the mass of the Sun and smaller. As a star forms, there is a lot of material that begins at large distances and collapses onto the stellar surface. If any of that material is initially rotating (which is probably true in all cases), then the rotation of the star speeds up as it collapses—the ice skater analogy again. For young stars, when they finish forming, their rotation is fast, often near the threshold where they would break apart if it were any faster.

If a star generates a magnetic field, which happens primarily with stars that are slightly more massive than the Sun or smaller, then its rotation slows down through a process called *magnetic braking*. This process involves the star's magnetic field and the constant stream of charged particles that the star sheds through the stellar wind. The charged particles in the wind are constrained by the magnetic field—they can move, but their motion is affected by the field. The magnetic field, in turn, is tied to the star itself—when the star rotates, its magnetic field rotates with it.

As the stellar wind streams away from the star, its connection to the magnetic field draws energy and angular momentum from the rotating star and imparts it to charged particles in the wind. Eventually, when these particles break away from the star's magnetic field, they carry that extra energy and angular momentum with them. The source of energy that they draw from is the rotational energy of the star—slowing the star's rotation. When a group of stars forms, their different rotation rates quickly (over the course of a few millions of years) settle onto a track where their spins slow down at a predictable rate, regardless of how fast or slow they were initially spinning.

The rate of the spin down depends upon the stellar mass: the more massive the star, the weaker its magnetic field, which is the essential component for this braking effect. A star's magnetic field is generated in a layer of convecting material near the surface. For the Sun, the layer constitutes about a quarter of its radius. The larger the star, the smaller this convective layer becomes, and the weaker the field. Eventually, if the star is massive enough, the convective layer disappears entirely, along with the bulk of its potential magnetic field. A weaker field means that magnetic braking is not as active and the star maintains a faster rotation rate. The most massive stars, which can't slow down because of their weak magnetic fields, often spin so fast that they are noticeably shaped like M&M candies—or *oblate spheroids*. Indeed, they are so distorted that the temperature near the stellar equator is cooler than near the poles—there is a belt of darker gas around their equators because the lower temperature material shines less brightly.

PLATE 1. William J. Borucki was the central figure in the development and operations of the Kepler mission. He served as the Principal Investigator for the mission. Source: NASA Ames.

PLATE 2. David G. Koch was the Deputy Principal Investigator for the Kepler mission and was responsible for completing some of the final technical demonstrations needed for the mission to proceed. Source: NASA Ames.

PLATE 3. Kepler's first-light image. This image shows the footprint of the Kepler photometer on the sky. It had fourfold symmetry (except for the two chips in the center) to continue to take data on most of the Kepler targets, even though the spacecraft itself needed to rotate, 90 degrees at a time, to keep the solar panels pointed toward the Sun. Source: NASA/Ames/J. Jenkins.

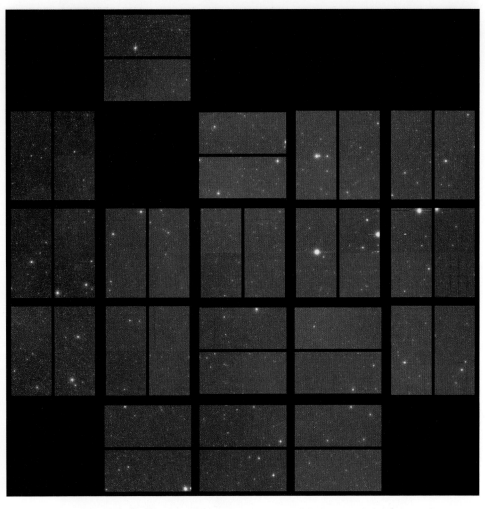

PLATE 4. Kepler's (K2) last-light image. Over time, different modules (pairs of CCD chips) failed. This shows the Kepler field of view at the end of its life. Source: NASA/Ames Research Center.

PLATE 5. Kepler-10. The first definitive detection of a rocky exoplanet. Kepler-10 is slightly larger than the Earth and orbits its host star in slightly less than one day. Source: NASA.

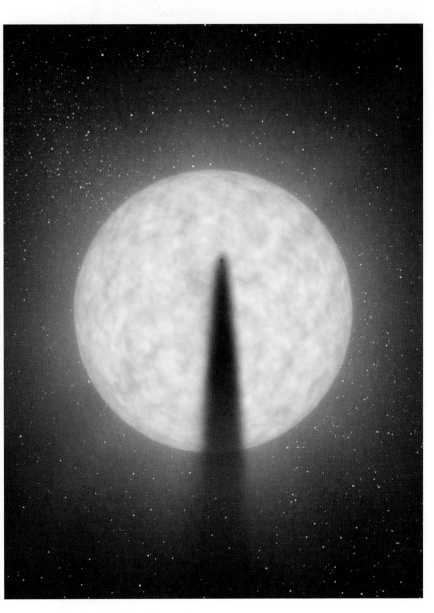

PLATE 6. KIC-1255 (Kepler-1520). This planet, with an orbit of only sixteen hours, is being vaporized by the intense radiation from its host star. The evaporated material trails behind the planet in a tail similar to a comet tail. Source: NASA/JPL-Caltech.

PLATE 7. Kepler-9. The first exoplanet system that showed multiple transiting planets and the first system that showed transit timing variations (TTVs) from the planet's mutual gravitational interactions. Source: NASA/Ames/JPL-Caltech.

PLATE 8. Kepler-11. A rich system of six planets orbiting a star like the Sun. The planets in Kepler-11 all orbit interior to the Earth and show transit timing variations (TTVs) from their mutual gravitational interactions. Source: NASA/T. Pyle.

PLATE 9. Kepler-16. The first circumbinary planet, a planet that orbits both stars in a binary pair, was detected by Kepler in 2012. Circumbinary planets are difficult to find in the Kepler data because the motions of the stars cause the transit shapes to differ from each other. Source: NASA/JPL-Caltech/T. Pyle.

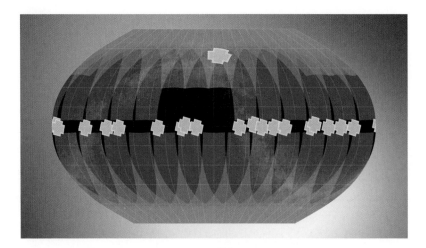

PLATE 10. Map of the Kepler, K2, and TESS fields of view. The primary Kepler field of view is near the top of the image, just above the galactic plane that curves through the image. The K2 campaigns all lie along the ecliptic, the plane of the Earth's orbit, which runs across the center of the image. The TESS fields of view overlap primarily at the ecliptic poles and form a flower-like structure, with each petal being observed for roughly one month. To avoid light scattering into some of the cameras, two TESS fields were shifted northward by one camera length. Source: NASA/JPL-Caltech/T. Pyle, J. Vargas (IPAC).

For stars like the Sun, the spin-down process continues throughout their lives as the rotational energy of the star is depleted by its magnetically coupled wind. As the star's spin slows, the effect gets weaker and weaker, but it persists as long as the magnetic field remains, the star spins, and particles boil off its surface. This changing rate of rotation allows us to estimate the age of the star by comparing its spin rate to the prediction from gyrochronology models—the models that Søren was working on with his study of the rotation of stars in clusters.

The Kepler field of view had four star clusters, each of a different age. Each cluster contained stars with a wide range of masses. With these clusters, we had four snapshots of how stars of all masses evolve over the time period spanning the ages of the youngest and oldest clusters. Using these data, we can calibrate the gyrochronology models to other measures of stellar ages. That, in turn, can be used to make more accurate age estimates for stars that are not part of the Kepler sample, or that are too dim to perform a full asteroseismic analysis [55].

Accurate age information has all sorts of implications for the history and future of planetary systems. Consider what we saw with the potential instability of the inner solar system. Given the short orbital periods of many of the Kepler planetary systems—orbits measured in days rather than months or years—the dynamical age of some of these planetary systems is mind-boggling. The Earth has completed about five billion orbits around the Sun, Jupiter about five hundred million. In contrast, some of the Kepler planets have orbited their host stars nearly five trillion times—a thousand times more than the number of orbits by the Earth. Dynamically speaking, as measured by the number of orbits by the planets, many Kepler systems are ancient.

There may be some processes that affect the orbital relationships among planets in their systems, but which occur on such long timescales that we haven't experienced them in the solar system. We know that the solar system is chaotic on billion-year timescales, with about a one-percent chance of the inner solar system going unstable in the next billion years. Now imagine giving the solar system a thousand tries to do just that—it virtually guarantees some radical changes. This is the scenario likely experienced by many Kepler planetary systems;

dynamical effects that manifest themselves on long timescales, effects that the solar system will never experience because the orbital periods of the planets are so much longer, have undoubtedly affected a number of Kepler systems. Exploring connections between the architectures of planetary systems and the ages of their host stars is work that is still ongoing.

With an astronomical survey the size of the Kepler data set, it became increasingly important to automate whatever work could be done. Measuring rotation periods for the stars was one example. Initially, the rotation periods would be measured one at a time. However, in 2014, Amy McQuillan (then at Tel Aviv University) led a study that measured the rotation periods of thirty-four thousand stars. Her results, in what became affectionately called (at least by some) the "shrimp diagram," due to its uncanny resemblance to a shrimp in profile, showed an interesting division in the stellar rotation rates [56]. Smaller stars seemed to cluster around two different rotation periods—spinning once in either roughly twenty days or roughly forty days. That division doesn't appear for more massive stars like the Sun or heavier. Something strange was going on for the smaller, M-dwarf stars. Figure 6.2 shows the shrimp diagram—the rotation periods of thirty thousand stars as a function of the mass of the star.

After nearly a decade of further study, an explanation is starting to materialize for this gap in the rotation rates seen in the shrimp diagram—though it remains unsettled. What we might be witnessing in the Kepler data, and in data from more recent missions, is the evolution of the spin of stars due to friction between their cores and the convective envelopes. The idea is that the cores initially rotate more rapidly than the envelope. But when stars are young there isn't a strong coupling between the two layers—they essentially slide past each other with very little friction. The outer layers of the stars slow due to magnetic braking. As the star evolves, the core is eventually able to connect more strongly with the envelope and to bring the two layers into a common rotation rate.

The slowdown of the spin is halted at the shorter rotation period (the faster spin of twenty days) because the fast-spinning core drags the envelope along, keeping its rotation fast and preventing the magnetic

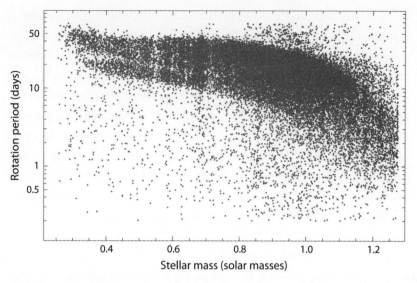

FIGURE 6.2. The rotation periods of thirty thousand Kepler stars as a function of the mass of the star. The gap that opens up for low-mass stars is likely due to the difference in rotation rates between the cores of the stars and their convection zones. When the core and convection zone are decoupled, the core rotates more quickly than the convection zone. As friction causes them to couple, the core speeds up the rotation of the convection zone (shortening the rotation period to a value near twenty days). After the core and convection zone are coupled, magnetic braking causes the star to rotate more slowly, near forty days.

braking from slowing the envelope further. Once the rotation of the core and envelope couple together, the star starts slowing down again as magnetic braking reasserts itself. The system makes up for lost time by quickly slowing to the longer rotation period, landing at a rotation period near forty days, at which point they continue to evolve as though nothing had happened. This transition to the longer rotation rate is fast by stellar standards. We can tell that the status quo reestablishes itself in short order because there are so few stars that have rotation periods between the two populations. That is, stars cross the gap between the fast- and slow-spinning states so quickly that we don't catch many in the act.

The same feature doesn't appear for higher-mass stars because, once the mass is near that of the Sun (roughly eighty percent of the Sun), the star develops a radiation zone in its interior. Rather than being a two-layer star with a core and a boiling, convecting envelope, these stars have three layers—the core, a radiation zone, and a convection zone. For the Sun, the radiation zone comprises a huge chunk of both the mass and the interior volume. The convection zone, while still substantial in size, contains only a fraction of the mass. The smaller convecting envelopes of the more massive stars are easier to drag along, so the coupling between the inner layers and outer layers happens sooner. The result is no pile-up of stars that are sorting out their internal dynamics, because they sort out quickly and at an early stage in their lives. I have to admit some degree of awe to think that we can actually see the effects of changing dynamics deep in the interiors of these distant stars just from the brightness variations that Kepler sees.

Another area of stellar astrophysics where Kepler shines is in understanding the details of multistar systems. Binary star systems are common, making up nearly half of the stars in the sky. But systems with more than two stars are also common—triple star systems being the next most abundant. There are a lot of triple systems in the Kepler data, most of which probably remain unstudied. (If you are going to spend a year or two of your life investigating something, you want to pick the most interesting thing you can find. This means that there remain a lot of slightly-less-than-the-most-interesting systems in the Kepler data that never rose high enough in the priority list to be studied.)

One of the most intriguing stellar systems was identified by Josh Carter (who also did the discovery paper for Kepler-36, the chaotic planet pair whose orbits are near the 7:6 ratio). He presented his early analysis of KOI-126 to the science team in the 2010 Denmark meeting. This system has two low-mass stars orbiting each other. That small binary pair, in turn, orbited a larger star roughly the size of the Sun. It turned out that the stars in the small stellar binary would eclipse each other, and together they would eclipse the primary star. The resulting brightness variations from this triply eclipsing system were rich in information about the orbital periods, inclinations, eccentricities, luminosities, and masses of the three objects.

This system was particularly compelling because we don't have many good measurements of the masses and sizes of small stars. The shortage of measurements makes it hard to calibrate our computer models for how those stars shine, what properties they have, and what their life cycles entail. So measuring the masses of the stars in the binary was important as a means to refine our models. The small stars in this system were among the least massive stars to ever be measured directly—having masses only twenty-one and twenty-four percent of the mass of the Sun. At the time the study occurred, there were only two other stars that also had radius and mass measurements of comparable quality (with uncertainties of only a few percent). Despite their diminutive sizes, such stars are among the most common in the galaxy, so by better understanding them we better understand the bulk of the stellar population.

As time went on, Kepler uncovered a number of similar systems with multiple stars orbiting each other, and generally having some of the stars eclipsing some of the others. Multistar systems (especially when they eclipse each other) have played a crucial role in our understanding of stars because they provide a means to measure their sizes and masses, as we just saw with the KOI-126 system. These quantities, along with their chemical composition, are the ingredients that we check for consistency with our computer models. The catalog of eclipsing binary stars from the mission eventually had some three thousand entries. From them we could measure the distribution of the orbital periods of the stellar pairs, telling us about both their formation and their subsequent dynamical evolution. For example, the tides that some stellar pairs raise on each other can be so extreme that their orbits decay to the point that the stars touch.

Since the members of these star systems often have different masses, we can learn how stars that formed at the same moment and from the same material evolve over time—similar to what we learn by studying star clusters. Computer models for how stars shine must produce the correct observable information for two stars whose only difference is their initial mass. Even better than characterizing these systems using dynamical measurements alone, we can estimate stellar properties from a combination of observables. For some systems, we can simultaneously measure the properties of the stars from their dynamics and from their asteroseismology signatures. Systems that are amenable to multiple avenues of investigation give a useful check on the predictions of both asteroseismology and dynamical models. In some cases, astronomers could see how the presence of a nearby stellar companion affected the pulsations of those stars.

Since roughly half of the stars that you see in the sky are actually multistar systems, an interesting question for Kepler is to find out the kinds of planetary systems that might exist in multistar systems. Exoplanet scientists had already discovered planets in binary star systems, but the stars in those systems were often far apart and the planets that we found would only be orbiting one of the stars. In some cases, the existence of the large distance to the second star didn't affect the orbits of the planets at all. In other cases it probably did have an effect, for example by inducing Kozai oscillations in the orbit of a giant planet and driving it into a three-day, hot-Jupiter orbit.

With Kepler, we found a binary star where each star had a planetary system. Kepler-132 (or KOI-284 as it was known while we worked on it) is a binary star with three transiting planets, where one of the stars has one planet and the other has two. There are likely additional planets orbiting both stars. In fact, since the initial planet candidates were identified, a fourth planet has been seen in the system, but the original three planets were what appeared in the early data. This system was a tough nut to crack, and it took some time to tease out what was happening.

When we first saw Kepler-132, we weren't sure what we were looking at. Two of the planets had orbital periods that were nearly identical, with one at about 6.2 days and another at 6.4. It is virtually impossible to have a pair of planets orbiting that close together without the system going unstable. The other planet candidate had a much larger orbital period and wasn't an issue in this regard. Aside from the fact that the two planets with nearly identical orbital periods couldn't exist in the same system, they passed all of the other diagnostic hurdles to make it into the candidate catalogs.

While we were figuring out how to grapple with this system, we modified a lot of the software we used for other studies specifically to exclude it, and any similarly close pair of planets. A high-resolution (speckle) image of Kepler-132 showed that the target star was actually a binary with two nearly identical Sun-like stars, which could not be resolved with the Kepler image alone. Steve Bryson, who developed the code to analyze the centroid shifts of the targets during transit in order to rule out background eclipsing binary stars, took special interest in Kepler-132. He found that the centroid of the Kepler data would shift in opposite directions during the transits of the two offending planet candidates with the nearly identical orbital periods. When one planet transited its host star, the center of the image would shift slightly toward the unobscured star, and vice versa. This result showed that there were planets orbiting both of the stars in the binary. This system was the first example of what we called a *split-multi*, or a multiplanet system where the planets were divided among different stars in a multistar system [57].

A final type of previously undetected planet in a binary star system is one where the planet orbits both stars at once—a *circumbinary* planet. Here, there is a binary star in the center of the system, and the planet orbiting the pair on a wider, exterior orbit. Kepler was the first instrument clearly capable of seeing a circumbinary planet if it transited across one or both of the stars. However, finding such a transit is easier said than done. We couldn't rely on the Kepler analysis pipeline to find one because, when the pipeline looks for a planet, it looks for dips in the light curve that are the same size, same shape, and that occur at regular intervals. With this approach, it is fairly easy to parse the data to look for a signal, verify that the signal repeats at regular intervals, and combine data from multiple transit events to boost the signal above any background noise. The downside is that if any of these assumptions are violated, such as the constant period or the consistent shape, the pipeline starts to fail. The changing orbital period from transit timing variations violates the constant period assumption, and this effect caused the Kepler pipeline to miss several planets because the transits didn't line up.

Circumbinary planets violate both of those assumptions. With each planet transit in a circumbinary system, it could be transiting one star, or the other, or both in some way. Because the stars orbit each other, their positions and speeds are constantly changing, and the planets can transit the stars at almost any point along the stellar orbits. If the planet happens to be moving in the same direction as the star that it is transiting, then the transit will take a long time to complete. If the planet is moving in the opposite direction, the transit will be shorter. The two stars could be eclipsing each other along with the planets, confusing the signal even more, and if the transit happens when the stars are eclipsing each other, then the shape of the transit will be different from when the stars are widely separated. It's a mess.

Because of all these issues, the initial search for circumbinary planets was done by eye—in an effort led by Laurance Doyle from the SETI Institute (SETI being an acronym for the Search for Extraterrestrial Intelligence). Laurance began with Kepler's catalog of eclipsing binary stars, and looked through those that had additional drops in brightness that the pipeline classified as threshold crossing events. He then combed through each system, looking at the light curves to pick out those where the extra transit-like signal looked like it could be caused by an additional orbiting body. One of the best candidates that he found was orbiting KIC 12644769—an eclipsing binary star with an orbital period of forty-two days.

Further scrutiny of this system showed three drops in brightness with separations of roughly 230 days and 220 days. Some object was transiting this binary in an orbit roughly aligned with the stellar orbits. The transits didn't occur at regular intervals, in part because of the dynamical interactions between the stars and the planet, which altered the orbital period of the planet, and in part because the stars were moving and weren't in the same relative position from one transit to the next. After ruling out all alternative explanations, the team determined that these transit signals were caused by a Saturn-sized planet, Kepler-16, orbiting the stellar binary.

The discovery of this transiting circumbinary planet was big news. Planets in systems with multiple Suns was the stuff of science fiction, like something from *Dr. Who*, or the planet Magrathea from *The Hitchhiker's Guide to the Galaxy*. Tatooine, the fictional homeworld of Luke and Anakin Skywalker, was a circumbinary planet as seen in the iconic image of Luke staring to the horizon as the two suns set in the distance. Some hints of circumbinary planets had been seen in the few years leading up to the 2011 announcement of Kepler-16, including one orbiting a binary pulsar. The planets orbiting binaries of Sun-like stars were inferred because of their effect on the stellar orbits—they did not transit, which made it difficult to definitively say that they were actual planets. Kepler-16 was a circumbinary planet where there was no question about its nature [58]. Plate 9 shows an artist's rendition of the Kepler-16 system.

The fact that Kepler-16 transited the stars allowed the team to make incredibly precise measurements of the sizes and masses of the stars, not just the planet. The smallest star, at only twenty percent of the mass of the Sun, had the most precise characterization of any star of comparable mass. Measuring the planet size and mass, its orbital inclination relative to the binary, and the eccentricities of the orbits of all the bodies, was enabled by the Kepler data. Over the course of the next half decade, nearly a dozen more circumbinary planets would be uncovered, most using Kepler data. Planet Hunters-1, discovered by a group of citizen scientists browsing through the Kepler data, is a circumbinary planet where the binary pair is actually part of a quadruple star system. And, in 2021, Jeremy Smallwood, a graduate student at the University of Nevada, Las Vegas, led a study of the first likely planet orbiting three stars at once (the planet's presence is inferred because of its effect on rings of dust surrounding the triple star) [59]. A circum-triple planet is sufficiently mind blowing that I only know of one such planet in science fiction, "New Texas" from the *Brave Starr* cartoon.

Circumbinary planets bring a new set of variables that can affect the properties of their orbiting planets, and have opened new avenues of scientific inquiry to understand the formation history and conditions for these planets. The amount of light that a circumbinary planet receives, for example, can vary significantly throughout its orbit—especially if the planet orbits in the same plane as the stars, so that the stars eclipse one another. On Earth, we can feel the difference on a summer day when the Sun dips behind a cloud. In a circumbinary system, one of the stars can completely block the light from the other—cutting the light that strikes a planet by as much as half. If you have a planet that is teetering on the edge of habitability, where liquid water might exist on the surface, the changing amount of light that the planet receives (its *insolation*) by this amount could push it over the edge, one way or the other.

Not only would the boundaries of a circumbinary habitable zone be affected by the constantly evolving distance between the stars and the orbiting planets, but the constantly evolving gravitational field from the stars would mean that planetary orbits that approach too near them will go unstable, ejecting the planets from the system or driving them into a collision course with one of the stars. Many of the orbits of circumbinary planets lie just outside that instability boundary, walking a fine line near the edge of their demise. The fact that so many live so close to that boundary suggests that there were likely many circumbinary planets that strayed too close to the stars and were forcibly removed from their systems. (When the planet-forming disk is still around, planetary orbits are more stable because the gas damps away their excited orbital eccentricity. So planets can initially form and survive closer to the binary pair. Once the disk dissipates, that protection is no longer available.)

Between the circumbinary planets that are seen to transit their host stars, and those that we've discovered because their gravitational influence affects the orbits of the stars, there is good reason to think that such planets are common. Roughly half of all stars are members of multistar systems. So multistar systems in our galaxy that could host planets probably number more than a hundred billion. Such systems clearly have a much different formation and evolutionary history from the planets in the solar system, and life in those systems would have to evolve in environments that are much more variable than what we experience on Earth.

Kepler's contributions to stellar astrophysics are unprecedented. The ability to peer into the interiors of stars, to measure the properties of the different layers, to see how they interact, how they affect orbiting planets and stars, and how they are (in turn) affected by them, constituted a giant leap for our understanding of these essential sources of light and energy in the universe. Measurements of their sizes and masses, how their spins evolve, and how they group together into multistar systems form crucial information for us to both know their inner workings and to appreciate the star that we orbit and that we see each day (except in Seattle, where it only comes out in the summer).

Our knowledge of the properties of extrasolar planets, and the conditions that might exist on their surfaces and in their atmospheres, is predicated on our knowledge of the stars that host them. We could not claim that a planet is truly Earth-like if we didn't know the environment created by its host star. The wealth of information Kepler provides in this area is an often unsung, but vital, addition to our understanding of distant exoplanets and how the solar system compares to them.

7

Planet Demographics

With our observations, and our analysis of those observations, we can determine the number of planets that a system has, where they are located, how they relate to each other, their sizes and orbital periods, their composition, and a variety of other properties. All of this information is fed back to astrophysics theorists, who use it to develop and refine models of the processes of planet formation and evolution. There is (or should be) an ongoing exchange between observers and theorists to see what is out there, explain what is out there, predict what else might be out there, and go looking for it. As the number of scientists in the field of exoplanets grew from a few dozen pioneers in the late 1980s and early 1990s to the few thousand that we have today, the sophistication of this exchange grew alongside—accompanied by a specialization of the roles.

To advance this enterprise takes much more than just observers who look through telescopes and theorists who tell them why they see what they see. It takes instrument makers to build more sensitive equipment that is tailored to the needs of the science, whether to mitigate the effects of the atmosphere or of temperature variations in the telescope, or to improve long-term stability and calibration of the devices. It takes statisticians to develop techniques to analyze the data that come from astronomical surveys, or to process signals coming from specific instruments, or to estimate contamination in the signal that might arise from unrelated astrophysical or Earth-bound sources. Theorists are engaged throughout the process, using known physical laws to predict the interactions of the light from the host star and planetary atmospheres and surfaces, or to model the motions of the objects in the system in order to rule out competing hypotheses that are proposed to explain the observations.

Eventually, the demographic information on the observed planets (sizes, orbits, etc.) works its way to the collection of theorists who work on high-level models of planet formation. These colleagues start their calculations at early stages in the process—perhaps with a simple cloud or disk of gas, or a field of debris and planetary embryos. They then explore the consequences of changing the properties of their starting conditions, or the effects of adding a new physical process that hadn't been incorporated into previous models. Ultimately, one of their goals is to match (or predict) the various characteristics of large swaths of the population of planets and planetary systems. However, in order to properly compare what we observe to what we model, we need to understand what we observe in the first place.

Consider the solar system. There are essentially three types of planets orbiting the Sun: rocky or terrestrial planets, gas giants, and ice giants. Then there are two large populations of debris, mostly in belts, one whose constituents are made of rock and the other of ice. The rocky planets being in the inner solar system is a consequence of the high temperatures that prevailed there during planet formation. It was too hot for light compounds like water or methane to condense, so the material that formed the planets was composed of minerals that freeze into solids at high temperatures—from several hundred to a couple of thousand degrees. The more volatile material could only condense beyond their respective *ice lines*, the imaginary line around the central star where the temperatures are cool enough to allow the compound to freeze. Near those ice lines, where the distance is smallest and orbital periods (that is, the dynamical timescales) are shorter, the planetary cores grew rapidly and began to pull material from the surrounding protoplanetary disk to form the gas giants. Farther away from the ice line, where the dynamical timescales are longer, the planets grew more slowly and formed the ice giants.

As we've continued to study the properties of the planets in the solar system, and tried to match those properties to our models, a number of discrepancies have arisen that have required refinements to the theory. One issue was the mass of Neptune. Our models kept showing that Neptune could not grow fast enough to reach its observed mass before the planet-forming disk dissipated. We know the typical lifetimes of protoplanetary disks since we can measure them around other Sun-like stars, and we can model the formation time for a planet like Neptune with a computer. The resolution to this issue was to have Neptune form closer to the Sun, where material was more abundant and the formation time was shorter. Then, once the disk was gone, Neptune's orbit could expand to its current size as the system of gas and ice-giant planets relaxed into a new equilibrium.

Evidence for this history appears in the orbits of the icy objects in the Kuiper belt—the belt of icy debris beyond Neptune. Many Kuiper belt objects were scattered into highly eccentric, and distant orbits, and a substantial number were captured into mean-motion resonances with Neptune, like Pluto. This model, called the *Nice model* (pronounced like "niece," and named after the city in France where scientists at the Observatoire de la Côte d'Azur pioneered its development), is accepted as the consensus theory for our origins—at least a consensus by astronomy standards [60].

A conundrum in the solar system similar to the large mass of Neptune is the small mass of Mars. Mars has only about one-tenth the mass of the Earth, yet our simulations for the formation of the rocky planets consistently produce planets at the location of Mars with masses that are much larger. Often, the masses predicted for Mars are larger than the Earth. The challenge of making such a wimpy Mars became so acute that, to explain the observations, scientists have proposed fairly drastic histories for the inner solar system. This issue has not been resolved, but one model that currently receives attention is called the *grand tack model*.

Here, at some point in the past, Jupiter's orbit began creeping toward the Sun—followed on its heels by Saturn. As Jupiter drifted inward, it scattered much of the planet-forming material in the vicinity of Mars and the asteroid belt, removing it from the area. Eventually, Saturn caught up with Jupiter and the two planets captured into a mean-motion resonance like Neptune did with Pluto. With their orbits now coupled, Jupiter and Saturn slowly retreated back toward the outer solar system. This saturnian intervention averted catastrophe for the Earth, which would never have formed had Jupiter continued its march unabated. For Mars, however, the damage had been done. With less material around to collect, Mars never grew into what it could have been.

Debate about the grand tack model continues, but it remains viable in the face of additional evidence. Not only does it explain the properties of Mars, but its predictions also match key features in the properties of the asteroids. Asteroids at different distances initially had slightly different compositions because of the temperature and pressure conditions that varied across that region of the solar system due to the changing proximity to the Sun. Today we know that some intermingling of the asteroids occurred that can't be explained if the planetary orbits remained as they are today. Their orbital eccentricities and inclinations also show signs of excitations in their past. The grand tack model provides the culprit for both of these observations.

Given the kinds of subtle insights into the history of the solar system that we can gain by examining the types of planets we have, and how they relate to each other and the Sun, knowledge of the types of planets that exist in exoplanetary systems and how they relate to each other can provide similar insights into the histories of distant planetary systems. We've seen this already with the formation of the hot-Jupiter and hot-Earth populations of planets. An important initiative for the science team was to glean the necessary insights from the Kepler data. This requires a careful accounting of the demographics of the exoplanet population—how many planets, of what type, live in which locations, and around what stars.

Among the challenges that we faced when gathering demographic information from the Kepler data was estimating how efficiently we were finding planets of different sizes that orbited at different distances. Finding one planet at a particular distance doesn't mean that there is only one planet at that distance. We needed to know how many planets we missed for each one that we found. Initially, the list of identified planet candidates, the catalog of Kepler Objects of Interest, was built by hand. Once the data went through the analysis pipeline, and the list of threshold crossing events that could be planets was tabulated, a group of scientists combed through the data, examining the light curves for each target to ensure that there weren't obvious problems with the data and to verify that the data looked sufficiently transit-like to warrant additional observations.

An issue with this approach is that people are different from each other—they are even different from themselves. Something that wouldn't have passed for a planetary transit at one point might look like one after several hours of constant work, or after a night's sleep. To address this issue with the human condition, scientists in the Kepler Science Office continued to develop and improve the analysis pipeline to make it more automated. (The science office was the group of NASA scientists who oversaw the scientific operations of the mission, like monitoring the spacecraft or running the data through the computers.) With each successive release of Kepler data, the production of the planet candidate catalogs relied more and more on computer algorithms rather than human intervention. Computer algorithms, while they may not be as good as a human at pattern recognition, are at least consistent—and consistency is important for estimating how well you can find a transit in the data.

The transition from humans to computers was envisioned from the start, but it took years to complete as new analysis methods were developed, then improved upon, and eventually incorporated into the software. You can't know up front exactly how best to treat the data, as there are always unknown sources of noise that don't appear until after you start collecting data and learning the personality of the instruments. Consequently, the pipeline development extended well beyond the lifetime of the flight operations. Officially, the Kepler mission ended in 2013; the final release of Kepler data was in 2016. But pipeline development persisted until 2018, when the final planet candidate list was published. This final catalog, led by Susan Thompson at the SETI

Institute, used a fully automated analysis pipeline. The catalog contained 8054 objects of interest and 4034 planet candidates. The planetary orbital periods ranged from six hours to nearly two years (632 days) and the sizes ranged from the size of the Moon to several times the size of Jupiter [61].

Bear in mind that just because the successive pipelines were more sophisticated doesn't mean that they were foolproof. With each catalog there were planets, often famous ones, that would disappear from the list of candidates and needed to be added back by hand. One example was Kepler-9. Kepler-9, the first system to be announced after the original slate of four hot Jupiters and one hot Neptune at the start of the mission, was the first truly novel system to come from the mission. It had multiple planets, we could see their dynamical interactions in the transit timing variations, and it had the hot-Earth planet orbiting close to the central star. The orbital variations, the primary reason we studied Kepler-9 in the first place, were large enough that the computer couldn't measure the orbital period correctly, or couldn't find the planets at all—so the system dropped off the candidate list, repeatedly. With each new candidate list, it was routine for our working group to email the science office asking them to add Kepler-9 (and a few other planets with similar issues) back to the list.

Despite these issues, four thousand planet candidates spread over such a wide range of distances and planet sizes was a treasure trove of information to be examined. Each catalog of new discoveries brought new questions about and new insights into these alien systems. One of the most important insights, indeed one of the primary science objectives of the mission, was to measure the planet occurrence rates—how many planets, what sizes, and what orbital periods. Thus, for each planet detection, we needed to determine how often similar planets were missed because they were either misaligned with our line of sight, or were overlooked due to noise.

The fact that most planets don't transit because their orbits don't properly align with our line of sight is an easy correction to make. We've also seen the work that went into eliminating false positives and noise sources. These are important corrections to make, but there was another important quantity to measure—the success rate, or *completeness*, of the analysis pipeline. The completeness tells us what fraction of the time the pipeline correctly identifies a planet of a particular size orbiting an arbitrary star in the data.

The completeness of the pipeline is influenced by all sorts of factors that have nothing to do with the planets themselves. A common example is planets with orbital periods that are near the orbital period of the spacecraft around the Sun. There are several environmental effects that repeat themselves over the course of each orbit, such as small temperature variations due to the slight eccentricity of Kepler's orbit or the angle of the spacecraft relative to the Sun. Since the pipeline looks specifically for repeating changes in the data, real planets can hide and imaginary planets can appear because of these seasonal variations.

The best way to measure the pipeline completeness is to test it directly by injecting fake planetary transits into the data and seeing whether the software recovers those planets. There were some discussions about running these tests in the meeting rooms and the hallways during early gatherings of the science team. In 2010, the science team lead, Ted Dunham (the same person who was instrumental in getting the participating scientists more active in the team), reached out to me and asked me to oversee the effort. He suggested that I work with a new staff member who had just joined the science office, Jessie Christiansen, to start formulating a plan to run these tests. I contacted her and we briefly discussed the idea. One of her primary responsibilities at the time was to prepare the documentation that was to accompany the release of the data—which was a shambles that required a lot of attention to streamline.

Unfortunately for the Kepler mission, Ted had also developed an instrument for another NASA project—the Stratospheric Observatory for Infrared Astronomy (SOFIA). This is a two-and-a-half-meter telescope that flies in the back of a 747 airplane. During flights, a large door opens in the side of the fuselage, allowing the telescope to peer out at its targets. The reason for such a theatrical observing platform is that SOFIA observes at wavelengths of light that are blocked by the atmosphere. Gathering data at those wavelengths requires getting above as much of the atmosphere as possible—either by putting the telescope in space, or (in this case) on the back of an airplane. SOFIA went airborne for science operations on May 26, 2010, right in the thick of the Kepler mission. Because of these competing demands, Ted had to devote more and more time to SOFIA, and less and less to Kepler. Over the next few months, the science team leadership transitioned to the deputy science team lead, Natalie Batalha.

The combination of the disruptions that came with changes in leadership and the fact that analyzing the Kepler data was already straining the available computational and human resources resulted in the task of assessing pipeline completeness being put on the back burner for some time. Fortunately, it wasn't forgotten and Jessie was able to plug away at the problem. She took the lead on what became one of the crucial series of tests of the Kepler pipeline and one of the most important scientific and technical results for the mission—required reading for anyone counting planets in the Kepler data. Her basic idea began with raw data from the spacecraft for a large number of target stars (it was actually the calibrated pixel data, where some known systematic effects and artifacts had been removed, but it was essentially raw). She then added computer-generated transit signals and false positive signals to each pixel in the images of the given target stars. In fact, she injected many signals into the data for the different target stars, hundreds of thousands of them.

The injected signals spanned a wide range of planet sizes and orbital periods. The effort was eventually split into two studies: The one led by Jessie was wide and shallow—meaning that injected synthetic signals were added to a wide range of target stars, usually one per star [62]. The second was led by Chris Burke and was narrow and deep, where a large variety of signals were injected into the data for a small number of stars [63]. The synthetic light curves were fed through the Kepler analysis pipeline to see what fraction of them the pipeline, and the other automated vetting software, could find. Once you have that information, you can make reliable estimates for the completeness of the Kepler planet candidate sample. That is, you will now know the fraction of planets that Kepler will successfully find as a function of the planet and stellar properties. The first study gives an overall view of the efficiency of the planet-hunting survey; the second tells you how the search for planets varies with stellar and planetary properties.[1]

The knowledge gained by these injection/recovery tests is essential to accurately measure the true occurrence rates of different planets. If you don't know how many planets your software will miss, you don't know how many planets each detection actually represents. If your software misses half of the planets that are in the data, then for every planet that you find, there are actually two planets that exist in the data. For large planets like Jupiter, the detection rate was well over ninety percent, and with the ongoing pipeline development it eventually reached a plateau of about ninety-five percent. As you look for smaller and smaller planets, where even small blips in the noise are sufficient to obscure a planet signal, the completeness of the catalog eventually dwindles to nothing.

You can't just run these tests on the early data and pretend that it will apply when more data become available. And since the pipeline software was constantly under development, you can't just pretend that the pipeline efficiency for detecting planets was the same from one version to the next. As the development of the analysis pipeline continued, the completeness calculations were updated to reflect those changes—squeezing every planet out of the data. With each new quarter of data, new batches of injection and recovery tests were needed—and many, many hours of both computer and scientist time. Early in the mission, time was in limited supply, especially since the science team members who were tasked with the occurrence rate measurements also had a mission to run.

Estimating the number of planets in the habitable zone of Sun-like stars was the primary objective of the Kepler mission. That estimate hinged upon the completeness of our analysis. So, while the official pipeline completeness work simmered, a similar test was being devised

---

1. One funny piece of knowledge that astronomers relearn from time to time is the fact that, when you need to transfer large amounts of data from one computer in one building to another computer in another building, the fastest way to do so is often to copy it to a hard drive and walk it over—as was done with Kepler data to run these tests.

by a graduate student at Berkeley—Erik Petigura. He wrote his own pipeline to find planetary signals in the Kepler data. While the two pipelines (Erik's and the pipeline from the science team) were somewhat different in their methods, the bulk of the planets found by them were the same. For the initial estimates of planet occurrence rates, finding most of the planets was good enough. So the study from Erik, estimating the occurrence rate of Earth-sized planets in the habitable zones of Sun-like stars, was the first one using injection/recovery tests to see the light of day. This work marked a significant milestone for the Kepler mission—it just so happened that it wasn't the official Kepler pipeline, but one that Erik developed specifically for this work [64].

His result was . . . complicated. A major issue facing estimates of habitable-zone planets is knowing how to define the habitable zone. There are many definitions, and many of those definitions are tenaciously defended by their advocates. They are, of course, all wrong— but that doesn't mean they aren't useful. (I'm quoting here a famous paper from 1978, where the statistician George Box titled one of the sections "All Models Are Wrong but Some Are Useful.") The utility of different habitable-zone definitions is that they all incorporate different effects that may be important for life to develop or persist on a planetary surface. The wrongfulness that they all share comes from the fact that habitability is a complicated question and there are undoubtedly things that are overlooked in any model. For example, in the solar system, the places we imagine looking for extraterrestrial life (Mars, Jupiter's moon Europa, Saturn's moon Titan, etc.) are all outside the habitable zone that exoplanetary scientists consider.

Having different definitions of the habitable zone means that scientists will make their estimates using the different definitions, which then makes it hard to compare the results of different studies. Those measurements may disagree with each other (and they often do), but for reasons that have little to do with the observations, and more to do with the selections they make based upon which definition for the habitable zone they use. Exoplanet scientists do try to steer away from definition-dependent results, but when we do so, that often means that we are reporting numbers that don't directly address the question that is central to the whole endeavor: How many habitable planets are there?

This brings us back to what Kepler actually measured. Erik analyzed the first fifteen quarters of Kepler data and found that the fraction of Sun-like stars that host Earth-sized planets with orbits of a few hundred days is around twenty percent. Erik's result happened to land near the middle of earlier estimates for this quantity. At the time, estimates of the fraction of Sun-like stars with Earth-sized planets in Earth-like orbits landed somewhere between none and a half (between one percent and forty-six percent). The wide range of possible values from the earlier measurements arose from different ways to estimate the completeness of the Kepler pipeline. In some cases there was no estimate for this quantity, others estimated the completeness by using measurements of eclipsing binary stars as a proxy for exoplanets—comparing the expected number of binary star systems in the Kepler data to the number of detected binary star systems.

To add some confusion, not all studies of the frequency of Earth-like planets measured the same quantity. Suppose there is a particular type of planet that you care about. To find these planets in the data, you first create criteria that must be satisfied for one to qualify. Then there are two common statistics that astronomers use to measure the number of those planets. Either they will estimate the frequency of stars that host your type of planet, or they will estimate the average number of such planets that orbit each star. These quantities are related, but are not identical. The former is essentially the probability that a given star will host a planet that meets the criteria. The latter gives the ratio of the number of qualifying planets to the number of stars—noting that it is possible for a star to host more than one acceptable planet. For example, with some small adjustments to the orbits of the planets in the solar system, we could imagine three Earth-like planets living in the Sun's habitable zone—Venus, Earth, and Mars. In this case, the Sun would add three planets to the count of the number of habitable-zone planets per star, but it would only add one star to the count of stars that host habitable-zone planets.

For measurements of either of these numbers, especially for Earth-like planets, there are large uncertainties that arise from two confounding issues. The first is that the sample of Earth-sized planets is small because their signals are difficult to pull from the noise—yielding fewer detections. The second issue is systematic uncertainties in the Kepler data—errors that arise from the instrument itself. For example, Kepler had an electronic noise source in some of the CCD chips. This noise was called the *rolling band* because it consisted of small ripples of brightness variations that drifted across the chips. If a star was observed by one of the affected CCDs, the rolling band would produce periodic dips in its light curve. Those dips could be dealt with easily within a given quarter, but after the spacecraft rotates, the star might land on a chip that didn't have that noise. Eventually, after four quarters, the star would land on the noisy chip again. At that point, it doesn't take much to trick the analysis pipeline into thinking there was a planet with an orbital period that roughly matched Kepler's orbital period of 372 days. As a consequence, there are many planet candidates with orbital periods near 372 days. If they were real planets, these signals would be right in the habitable zone of Sun-like stars. However, they are actually false positives that cloud our ability to estimate the number of real planets at similar locations.

Despite nearly a dozen independent attempts to measure the occurrence rate of Earth-like planets, agreement between the different studies has not been forthcoming. Groups of scientists on an experiment often develop separate algorithms to find signals in some scientific data. Having multiple studies is often done specifically with the intent to find discrepancies between different approaches so that errors in the underlying methodologies can be found and addressed. In my experience, different teams are often relieved when the discrepancies are relatively large and when they show up right away, since it means they are easier to find and resolve. If the differences in the results of two different methods are tiny, it can take a long time to ensure that you aren't missing something important.

Such has not been the case with occurrence-rate measurements with Kepler data. The challenge comes from the mix of the small number of detections, the large number of false positives, the systematic effects from the instrument, and our uncertainty in the properties of the planets and the host stars. (Recall that Kepler actually measures the size of the planet relative to the size of the star, so uncertainties in the star lead to uncertainties in the planet.) The most comprehensive attempt to measure this quantity came from Steve Bryson at NASA Ames and his collaborators.

Even this group, with a decade of experience working with the Kepler data, was frustrated by the difficulty in making this measurement. What they found at the end of their effort was that the frequency of planets with orbital periods comparable to the Earth, orbiting stars like the Sun, is about one in ten. But, given the uncertainties described above, and different ways to handle them, the range of possible values for the frequency of these planets extends from only one percent up to two hundred percent (that is, two such planets per star). There is clearly work yet to do to refine this number, and Steve and his collaborators are still chipping away at the problem as I am writing [65]. One fact that can be stated with some confidence, however, is that when you add up all types of planets (not just Earth-sized), and all types of stars (not just Sun-like), planets outnumber the stars in our galaxy.

We've also found that, at least when it comes to planets with orbital periods less than a few hundred days, smaller stars have more planets than their larger counterparts. Since the majority of stars in the galaxy are smaller than the Sun, this observation implies that small planets orbiting small stars may be the most abundant type of terrestrial planet. Many of these planets likely orbit within the habitable zones of their host stars, which suggests that most life in the galaxy may also reside on these small planets. However, while potentially habitable, these planets may not be very hospitable.

Tidal deformation of the planets would cause the spin of the planets to couple to their orbits—as we've seen with Mercury and with many of the moons in the solar system. However, the presence of other planets in the system will cause long-term variations in the eccentricity, and even the period of the planetary orbits. Eccentricity variations would disrupt the spin–orbit coupling of the planet because the planet would try to keep the same face toward the host star at the point of the planet's closest approach. If that distance keeps changing because of the changing eccentricity of the planet, then the spin of the planet would occasionally decouple from its orbit—the planet's spin can't change fast enough to track the changes in the orbital shape. Orbital period changes caused by neighboring planets would also decouple the spin and orbit.

The result of this decoupling, from the perspective of someone standing on the planet, would be the planet's rotation randomly changing direction in a chaotic fashion (real, mathematical chaos). For a few days, the Sun would rise in the east and set in the west, then it would hover in the sky, moving back and forth from one horizon to the other—half the time you would be on the daylight side, and half the time the sky would be dark as the Sun hovered over the opposite side of the planet. After a few, or many, or thousands of orbits, the Sun would start rising again in the west and setting in the east. A constantly shifting daylight pattern that could plunge your residence into a millennium of darkness is not conducive to life beyond simple organisms. There would be no consistent day–night cycle, no consistent weather patterns, and a constantly changing climate.

We've seen how the orbital periods of planets in a system, especially if they are near mean-motion resonances, tell us about their dynamical histories. The relative sizes of planets in the system can give complementary information. In the solar system, the terrestrial planets are similar in size (as determined by their radii), with Venus being nearly identical to the Earth, while Mercury and Mars are one-third and one-half the size of Earth respectively. The ice-giant planets are nearly the same size as each other, at four times the size of Earth, and the gas giants are about ten times the size of Earth. These differences give clues to the differences in their composition and the processes of their formation.

The majority of planets in Kepler planetary systems are between the size of Earth and the size of Neptune. They fill a space in the distribution of planet sizes with no counterparts in the solar system. Beyond the size of Neptune, the data show a steep decline in the number of exoplanets. There are a few planets larger than Neptune or Uranus, hot Jupiters for example, but these are relatively rare compared to their smaller counterparts—one percent for hot Jupiters, and a few percent for everything else that is larger than Neptune.

The peak in the distribution of planet sizes seen by Kepler is near two Earth radii. Most of these planets have orbits of a few tens of days. These planets are unheard of in the solar system, where no planets exist at that size, nor do any have such short orbits. Measurements of their masses place most of them at a few to several times the mass of Earth (three to sevenish). Given these properties, it remains a mystery how these planets form. Moreover, given that they orbit a sizable fraction of stars in the galaxy, we aren't really sure why the solar system didn't turn out like them.

An early study by the science team examined one clue that might help answer this question. David Ciardi, an astronomer at the NASA Exoplanet Science Institute, led an effort to examine the relative sizes of sibling planets in multiplanet systems. He and his team found that planets in a system are often roughly the same size, but those that are farther away from the host star in a given system are slightly larger than those that are closer [66]. These are averages, so there remains substantial variation—individual systems can have planets that are five or ten times larger or smaller than their interior neighbors. But weird ones, like Kepler-36 with its rocky interior planet and nearby but fluffy exterior planet, or WASP-47 with its hot Jupiter surrounded by smaller, terrestrial worlds, while rare will often receive more attention than the less remarkable but more common planetary systems. A typical planet pair within a typical system is a bit more likely to have the larger planet on the larger orbit. This result may have implications for the amount of condensed material that was available to form planets at greater distances, or it may be a consequence of processes that occurred after the planets formed, or both.

It's important to bear in mind that the majority of Kepler planets have orbits that are less than one year—orbiting what would be well interior to Earth's orbit. So Kepler is gathering data for what is only a fraction of all planetary systems, and likely for only a fraction of the planets within those systems. In the solar system, Neptune orbits at a distance thirty times that of the Earth, so there is plenty of unexplored real estate in distant planetary systems. It is possible, even likely, that we would see abrupt changes in planet sizes at certain distances, or around certain stars, if we had a larger sample of planets in larger orbits, such as orbits that are outside the ice line in their respective systems, where the gas and ice giants of the solar system roam. Nevertheless, for the inner solar system, both of these trends show up. All of the terrestrial planets in the solar system are within a factor of two of their average size. And Earth is larger than Venus, which is larger than Mercury. Mars doesn't follow this pattern, but not every planet in the Kepler data follows the pattern either.

This result was expanded in a later study, led by Lauren Weiss (now at Notre Dame, but who started working on Kepler-related science as a graduate student at Berkeley). Lauren and her collaborators showed that, in most circumstances, all planets in a system are similar to each other in size [67]. So, if one planet in a system is relatively small, the other planets in that system also tend to be small. If one planet in the system is large, then the other planets in that system will tend to be large—like "peas in a pod" to steal her description of the phenomenon. This finding may suggest that the environment in the disk where a set of planets form changes little across a wide portion of that disk. A disk with material of a certain density and composition will produce planets of a given size. If the density or composition of the disk changes, the sizes of all of the planets it produces will also change. This is likely true regardless of the details of the formation process for small planets. By contrast, rare is the system where one planet is wildly different from a neighbor—not unheard of, and definitely interesting, but rare.

As we look at planets that orbit closer to their host star, there is one notable change that appears. We see in the Kepler data a transition between two different kinds of small planets. This transition doesn't happen all at once, nor does it happen for all systems, but as you get closer to the star, the planets get smaller—and their sizes cluster into two distinguishing groups. We generally interpret this transition as one between planets that have substantial atmospheres (which orbit slightly farther from the central stars) to planets with little or no atmosphere in closer orbits. Some recent work suggests that, rather than an atmosphere being the primary difference, it is the presence of a lot of water on the planet that makes the planet larger and less dense. This isn't wimpy amounts of water like on the Earth, we're talking a third or a half of the planet mass being made of water.

It is surprising that this transition between these two types of planets is so easy to see. Planets less than about 1.8 times the size of the Earth appear to be made primarily of rocks and metals. Planets larger than this value have larger atmospheres or more water, which inflates their sizes while adding relatively little to their mass. There are only a few planets in the middle ground that lies between these two different populations. This lack of planets in the transition between the two populations in planet size does not have an obvious explanation. In principle, there should be smaller planets with moderately thick atmospheres, or larger planets with unusually small atmospheres, but the group of planets that lie in the range between the two types is small.

There were some early hints at this gap between the more and less dense planets—especially when looking at single-planet systems. However, in order to make a definitive discovery, we needed more precise measurements of the planet sizes. That meant we needed improved measurements of the sizes of the host stars, since it is the ratio that the Kepler data provide. If the uncertainty in the size of the host star is large, then the uncertainty in the planet sizes will be similarly large, smearing out the observed distribution of planet sizes and obscuring the presence of the two populations. More precise measurements of the properties of the host stars were essential to tease out the feature.

Lauren Weiss did a study of Kepler planets in 2014 where she used the Keck telescope both to measure planetary masses through their Doppler signal, and to make improved measurements of the sizes and masses of the host stars. In this early work, she estimated that the transition to rocky planethood occurred for planets smaller than one and a half times the size of the Earth—near the value we accept today. Later, a group of astronomers (including Lauren) from Caltech, Berkeley, and Hawaii, and led by a graduate student Benjamin Fulton, made the definitive measurement of this *radius valley*, or what is occasionally called the *Fulton Gap* [68, 69].

Theorists are still working to explain the origins of this radius valley. An important clue is that the divide between the more dense planets and the less dense ones occurs at slightly larger sizes when the planets are nearer the host star (the relationship is weak, but seems to be real). Ideally, an explanation should account not only for the existence of the gap, but also for this slight dependence on the orbital distance. One of the more promising explanations is that the amount of atmosphere that planets have, and hence the size of the planets when the atmosphere is included, was initially more similar across all planet masses. But high-energy radiation from young stars (which are often very active) started evaporating the atmospheres of the nearby planets. If the dense material of the planet core was sufficiently massive, it could retain the atmosphere, or at least hold onto it until the star calmed down. By contrast, if the planet core wasn't massive enough, and its gravitational pull was too weak, the atmosphere would evaporate almost completely—leaving behind only the rocky material. This hypothesis explains the two different kinds of planets, and shows why planets that are closer to the host star are more frequently on the rocky side of the gap. The radiation from the active star is more intense nearer the source.

An alternative explanation for the abrupt transition from planets with thick atmospheres to those without is that the energy to dissipate the atmosphere may come from the planet itself. Planetary cores initially grow by accreting rocky and metallic materials onto their surfaces. This is not a slow, peaceful process. Space rocks frequently hit the Earth with speeds over fifty thousand miles per hour. Exoplanets would experience the same bombardment, and the energy of those impacts would heat up the planetary interior. Once the pummeling stops, the planet core (meaning all of the solid stuff enshrouded by the atmosphere) would cool off. It would do this by transferring that energy to the atmosphere, which boils off the planet provided that the planet is too small to keep it trapped by gravity (which is easiest to do if it is made from low-mass elements like hydrogen and helium).

This core-driven mass loss would be an effect that lasts longer than mass loss from stellar irradiation. The Earth is still cooling from its formation—though at a rate that is too slow to evaporate our atmosphere. Cooling exoplanets can continue to shed their atmospheres for a few hundred million years. By contrast, the high-intensity radiation from early stellar activity might only last for ten million years. The truth might be a combination of both explanations. As we study the properties of planets orbiting young stars, we may be able to determine the relative importance of these two processes in shaping the population of planet sizes.

Still another explanation for the different planet sizes is giant impacts. In the latest stages of terrestrial planet formation, there are a fair number of large, Mars-ish-sized objects circling around the star. These are the final objects to merge with the planetary embryos on their path to making something like Earth. When they hit, they pack a punch. A widely accepted account of the Earth's history is a collision between the proto-Earth and the Mars-sized Theia. In one version of this story, the collision nearly destroyed the Earth, causing it to spin so violently that it formed a *synestia*, a nearly donut-shaped blob of molten rock. Portions of that blob eventually coalesced to form the Moon, while the remainder settled back down to become the Earth. Collisions like these have sufficient energy to strip off a significant amount of light gases, and similar collisions in exoplanetary systems could readily remove a large, but tenuous atmosphere from a planet.

If the observed gap in planet sizes is due to the presence of large amounts of water, rather than differing amounts of atmosphere, it may imply that the planets formed in different parts of the protoplanetary disk. Had they formed near or beyond the ice line, more water could be trapped in the planet—yielding the observed larger size. Given that the two planet sizes appeared at similar orbital distances, water-rich planets would likely have migrated inward from the more distant, cooler parts of the disk. As we've seen, a record of that migration history might show up in the orbital period ratios of the planets. Investigations into the consequences of each of these mechanisms for forming a gap in planet sizes are ongoing. More data, especially with younger stars and younger planetary systems, should help disentangle these explanations and give us a better understanding of the history of the whole population of planetary systems.

Ever since the Kepler pipeline started producing its rapidly expanding list of planet candidates, the race was on to learn everything we could about them. The Kepler Follow-up Observing Program (KFOP) was a large consortium of ground-based telescopes and observers tasked with doing the necessary work to confirm and characterize the planet candidates. As we've seen, a portion of this work was verifying that there were no interloping stars that would cause a misestimate of the planet size or false positive signals from distant eclipsing binary systems. Considerable time and effort also went into confirming the planetary nature of the transit signals—showing through Doppler measurements that the orbiting objects had low-enough masses to be considered planets (as opposed to small stars, which can often be similar in size to Jupiter and Saturn, but with a hundred times their mass).

Another important set of measurements for our ground-based telescopes was to verify the basic properties of the host stars—especially confirming that the stars were of the expected type (main-sequence, Sun-like stars or M-dwarf stars as the case may be). When the Kepler Input Catalog was assembled prior to Kepler's launch, it was built by classifying stars based upon their "colors." These are *photometric* measurements of the star, where the properties are determined by how bright the stars appear in several wide bands of the electromagnetic spectrum (blue, red, green, etc.). Comparing the brightness of the target stars in these bands gave a rough estimate for stellar properties like the stellar temperature and surface gravity.

The advantage of photometric measurements is that you can roughly characterize hundreds or thousands of stars at a time. However, photometric measurements have large uncertainties, limiting our understanding of both the star and its orbiting planets. By contrast, *spectroscopic* measurements, where you spread the light out into all of the colors, give much-higher-resolution measurements of these properties, but you can only make those measurements for one, or a few, stars at a time.

Once Kepler started finding planets, we no longer needed to survey tens of thousands of stars—the observers could concentrate on stars with known planets. With the list of planet-hosting stars in hand, the follow-up program, and several groups outside the science team, began measuring the properties of the host stars in better detail. This work allowed us to make important measurements about planetary demographics—the frequencies of different planet sizes and orbital distances of planets—which were primary motivations for the mission.

One of the first investigations into the big-picture planet occurrence rates came in the middle of 2011 and was led by Andrew Howard (now at Caltech). The first study only considered planets with orbital periods shorter than fifty days [70]. That result was a huge deal. Courtney Dressing, who was working with the group at Harvard, recalls receiving a file in her mailbox labeled "Top Secret: Eyes Only," showing these results. The advice from her advisor (David Charbonneau) was to try to reproduce the result for the smaller M-dwarf stars [71]. That work, and subsequent studies throughout the 2010s, extended Andrew's results to larger orbits and to other stars.

Over the course of several years, the stellar measurements became more precise. A significant advancement came from the GAIA mission from the European Space Agency. GAIA made high-precision distance measurements to far away stars by triangulating their positions using the orbit of the Earth as a baseline. Its observations improved our estimates for several stellar properties, especially their sizes. This, in turn, improved our planet size estimates and enabled new discoveries about planetary demographics, such as the improved characterization of the radius valley mentioned earlier. It also gave us insights into how planetary systems differ as a function of the type of star that they orbit.

After more than a decade of measurements from different instruments, different surveys, and different planet-detection techniques, we have a reasonable consensus of what planetary systems look like, at least for planets that are currently detectable with our equipment. The information we get from Kepler applies primarily to planets that orbit out to half the distance of Mercury for the smallest stars, and out to the orbit of Earth for larger stars. From Kepler alone we find that there are more planets than stars. Most planets that are interior to the Earth's orbit are smaller than Neptune—divided between some planets that have modest atmospheres (or a lot of water) and some that have thin or no atmospheres, with a measurable gap in size between the two types. Around half of the stars in the sky host planetary systems like the ones Kepler found. For the other half, we don't really know—they may have virtually no planets, or they may all look similar to the solar system (which Kepler doesn't have a good handle on because they are right at the limits of its detection capabilities).

The abundance of small, super-Earth and sub-Neptune planets rises steeply as a function of the planet's orbital period, from a few percent of stars having planets with orbits of two days or less, up to several percent that have planets with orbits of about ten days. For larger orbits, small planets are equally likely to have orbits of ten days as they are to have orbits of a year (beyond which the Kepler data provide little information because of the limited duration of the mission). These small planets seem to be more common around small stars than they are around large stars, noting that there remains some ambiguity because they are easier to detect around the smaller stars.

The abundances of small planets don't change much if you change the composition of the material from which they form. We infer this trend from the composition of the host stars. When compared to the abundances of heavy elements in their host stars, the frequency of small planets changes very little. The same is not true for giant planets, which prefer stars with lots of extra elements beyond helium. Ultimately, we estimate that Earth-sized planets in Earth-like orbits probably orbit roughly a quarter of all stars.

There is a small cluster of giant planets like Jupiter and Saturn at three-day orbital periods (about one percent of stars have hot Jupiters), but these planets are more rare for orbital periods of a few weeks. They become increasingly common as the orbital periods become longer. When Kepler data are combined with the Doppler surveys, we see that the giant planet population peaks at orbital periods of around fifteen hundred days, a bit interior to the orbit of Jupiter, but at a place where the formation conditions would be similar to what Jupiter would have experienced. The abundance of giant planets then tapers off for larger orbits. Unlike the more abundant, smaller planets, gas giants only orbit roughly one in five stars. It isn't clear at this time whether the Kepler-like planetary systems preferentially have (or lack) gas giants in these wider orbits, or whether there is no correlation between the presence of outer giants and inner super-Earths and sub-Neptunes.

We know that gas giants can have significant effects on the orbital properties of interior, small planets. We see this with the chaotic behavior of Mercury's orbit in our own solar system. In many ways, small planets survive at the whims of the giants, so I would be surprised if there wasn't some noticeable consequence of the presence of gas-giant planets on their inner siblings. Nevertheless, because the Doppler surveys (which find the long-period giant planets) and transit surveys (which find the short-period transiting planets) have different reasons for selecting the stars that they study, the correlation between their results isn't firmly established—more data that combine different detection methods will tell us more about whole-system architectures.

Finally, from gravitational microlensing and from direct imaging surveys, we have some handle on the planets living in places that are hard to reach for transit and Doppler surveys. We are still in the early stages of this work and can expect more results from all of these detection methods in the coming years. Directly imaged planetary systems indicate that about five percent of stars have Jupiter-mass planets on wide orbits—really wide orbits. These orbits can be a hundred times larger than the orbit of the Earth (or more), which is more than three times the orbital distance of Neptune around the Sun. We see additional evidence for large planets at these distances around stars that are still forming because the forming giant planets produce gaps, wakes, and vortices in the disks where they reside. These features appear in images taken with large arrays of telescopes designed to look at them—for example, the Atacama Large Millimeter Array (ALMA).

Gravitational microlensing results reveal a large population of Neptune analogs orbiting distant stars. A large, future microlensing survey will be part of the anticipated Nance Grace Roman Space Telescope, a planned flagship mission from NASA. Like Kepler, the Roman telescope (originally called the Wide Field Infrared Survey Telescope, or WFIRST) had been cooking for several years before it was chosen to move forward. This project received a shot in the arm with a donation to NASA of a pair of large, high-quality, wide-angle mirrors by the US National Reconnaissance Office.

What the US intelligence agencies were doing with a pair of large, high-quality, wide-angle space telescope mirrors is anybody's guess. With Roman, we know that at least one of the two mirrors will be looking at Earths that are orbiting stars other than the Sun. In the coming decade, Roman's microlensing survey should improve upon our current microlensing results and tell us how frequently sub-Neptunes and super-Earths appear in wide, multiyear orbits—beyond what Kepler was able to see.

8

Kepler Shows Its Age

The Kepler spacecraft started showing signs of age relatively early in its life. Space is a harsh environment. Away from the Earth's magnetic field the telescope is exposed to the charged particles streaming from the Sun in the solar wind. Without the presence of Earth's atmosphere, it feels the brunt of high-energy cosmic rays released in stellar explosions or the distant nuclei of active galaxies. The solar system is filled with dust particles and micrometeorites. If these last items (dust particles and micrometeorites) don't sound very intimidating, imagine a grain of sand being blown into your face by a 10,000 mile-per-hour wind. That is the typical speed for objects passing by each other in the inner solar system.

All of these sources, and more, start chipping away at the health of the spacecraft and its instruments. As the cosmic rays pummel the CCD chips, they slowly damage the components and degrade their quality. Over time, the camera is less and less capable of detecting what it set out to do. This fact, of course, was known to the designers of the mission and it was a consideration for the specifications of the affected parts. Even when these parts degrade, many of the effects of the environment can be mitigated by changing how the data are analyzed after they are taken. For example, there are a lot of tools to help scientists correct for artifacts in the data from the Hubble Space Telescope (anomalous signals from worn parts or other peculiar events) that are caused by space-related weathering.

In addition to the slow decomposition of detectors, surfaces, and other stationary parts, telescopes often have a number of moving parts that wear out as they are used. For Kepler, the most important of the moving parts were the reaction wheels. These are fast-spinning wheels used to maintain a stable attitude. Without the stabilization that they provide, it is not possible to make accurate observations of the designated field of view. As the telescope drifts, or is impacted by debris, or has just finished its turn toward the Earth to downlink the data, these reaction wheels adjust their rotation speed to effect tiny adjustments in the pointing of the telescope. If the reaction wheels fail, the telescope-pointing capabilities decline, which introduces noise that obscures the planetary transit signals.

Despite the inevitable decline in the instrumentation and the moving parts, these do not always pose the fundamental limit on how long the satellite will last. For the Spitzer Space Telescope, the limitation for some of its instruments was coolant. Spitzer was an infrared telescope, observing long-wavelength light, such as that emitted by warm objects like people or infrared space telescopes that are basking in sunlight. Spitzer launched with a supply of liquid helium on board, which has a temperature of $-270°C$, or $-450°F$. The instruments were kept cool by evaporating away that helium supply. After the coolant boiled off, the instrument started to warm until eventually the telescope glowed more brightly in those wavelengths than the objects astronomers would otherwise like to observe. Two of the three instruments were shut down completely, as were two of the four passbands on the remaining instrument (the detectors that observed at the longest wavelengths, which were most affected). The telescope continued operations for another decade during the "warm Spitzer" phase, until it was eventually shut down in 2020.

Many NASA missions can operate far past their warranty. The Mars rovers Spirit and Opportunity were slated to operate for about 93 Earth days (or 90 martian days). Instead, Spirit lasted for 2200 days and Opportunity for 5200, or more than six years and fourteen years respectively. For Kepler, the fundamental limit to the spacecraft lifetime was set by its onboard supply of hydrazine—a propellant used in thrusters to steer the spacecraft. The moment its hydrazine ran out, the spacecraft would be left adrift with no way to recover. The nominal mission was three and a half years, but Kepler's ultimate lifetime depended on how efficiently it could use its fuel as it operated, and how much it would consume as it maneuvered its way into its initial observing position.

Following the Kepler launch, and once the spacecraft was situated, the flight engineers could estimate how long the reserves could last under standard operations. At a science team meeting shortly after launch, David Koch, the Deputy Principal Investigator for the mission, excitedly told me that the remaining fuel should last for eight years, and with efficient use it could stretch as far as ten. That was great news. It meant that Kepler could search for planets for more than twice the duration of the nominal mission, finding more, and longer-period, planets, as well as making important measurements for stellar properties and planetary system dynamics that stretched over a longer time interval.

Despite the available fuel, and the rosy initial prospects for a long mission, Kepler would have no such luck. Between May and December of Kepler's first year there were three events that drove it into safe mode. Safe modes occur when some sensor on the satellite trips and forces the instrument to turn off, protecting it from damaging itself if it were to continue operating with something significantly wrong. The triggering events could be anything from a response to cosmic rays, to some part of the spacecraft getting too hot, or just that the spacecraft gets confused by conflicting commands from the software. Once it is in safe mode, it stays there until the ground crew can figure out what happened and revive it. In the meantime, it stops collecting data—which is the whole point of flying the mission. Resolving safe modes quickly is a high priority.

While it is normal, especially at the beginning of a mission, for a spacecraft to enter safe mode, the events were topics of serious discussions during our science team meetings and phone calls. Not only would we lose data while the safe mode was ongoing, but because some devices would turn off, the instrument would change temperature, and a number of other small effects would temporarily degrade the quality of the data after the spacecraft had "recovered." Since data from Kepler are in the form of a time series, with ostensibly continuous observations, you can't simply make up for missing data by turning the camera back on—it would require traveling back in time to catch what you missed.

Ensuring that the spacecraft only enters safe mode for valid reasons, and that when it does we only turn off the most vulnerable equipment, is crucial for getting the most science from the mission. To put a crude number on the value of lost data, the mission itself cost around six hundred million dollars, and during its normal operations it would have roughly sixty thousand cadences. So each exposure is worth (monetarily) about ten thousand dollars. Having the spacecraft turn off for a few days, just because it got confused by a software bug, is not something you want to just live with if there is a way to fix it.

During that same six-month period of Kepler's first year, there were also three times when the telescope lost *fine point*, making it unable to maintain its orientation with the stability needed to take science-quality data. To properly point the spacecraft, and to keep the stars in its field of view landing on the correct parts of the camera, the spacecraft needs to know where it is looking, and to be able to track changes in its positioning. If the spacecraft attitude gets too far from the nominal direction, sensitivity variations within individual camera pixels and across different pixels could add a false planet-like signal or add noise that would obfuscate a real signal.

To maintain fine point, the stability needed for science operations, the spacecraft attitude had to stay within 9 milliarcseconds of the target direction. That is a tiny number. Suppose Kepler were located at its operations headquarters at the NASA Ames Research Center, near San Francisco, and it was pointed at the flag atop the Washington monument in Washington DC some 3200 kilometers (2000 miles) away. If it started to drift downward, scanning along the height of the monument a continent away, it would be out of the fine point specification before the pointing direction reached the monument's base, only 150 meters below the top. Or, if you prefer another example, if you were watching a bluebird with binoculars 100 yards away and staring straight into its eye, you would be out of fine point by the time you were staring at its tear duct.

To maintain this rather remarkable pointing stability, Kepler uses a set of star trackers. These are small cameras that image the sky and compare the stars that it observes to a known star map—making small corrections with the reaction wheels or thrusters to keep the image lined up with the star map. At the time the spacecraft was constructed, these star trackers weren't capable of delivering the needed pointing stability on their own. This mode of operation, using star trackers, is the *coarse-point* mode. While the telescope is aligned in approximately the right direction (to within a minuscule fraction of a degree), the spacecraft would shift from coarse-point to fine-point mode.

In fine point, it used four *fine guidance sensors* to maintain its attitude. The fine guidance sensors were small digital cameras wedged into the corners of the Kepler focal plane—filling a portion of the gaps left by the main camera. (The whole footprint of the focal plane is a five-by-five grid with each of the corners removed. The fine guidance sensors occupy a small piece of those corners.) These fine guidance sensors were more sensitive than the star trackers, and each monitored the positions of about ten stars—for a total of forty. They image the stars ten times per second, and send signals to the reaction wheels to correct for any movement of their images.

If the telescope drifts too far from the intended orientation, and loses fine point, the changes to where the light strikes the camera and the difference in the sensitivity across the individual pixels make brightness fluctuations that are too large to see Earth-sized planets. The data are no longer considered science quality. The 9-milliarcsecond cutoff threshold corresponds to a shift of two-thousandths of a pixel in the locations of the stars. The fact that a deviation so small renders the data unusable is a testament to the quality of data that Kepler produced. Over the course of four years, Kepler went into safe mode eleven times and lost fine point seven times—including the final time from which it never recovered. For ten months after launch, and eight months after the start of science operations (aside from the fine-point losses and safe modes) the telescope performed admirably.

Every 1765.5 seconds from the time of its launch, Kepler tabulated the brightness of the stars on its target list—each cadence timed to an accuracy of 50 milliseconds or better. Maintaining this 50-millisecond accuracy is no small feat. It isn't like you can just check the time using a stopwatch or the GPS satellites. Kepler is in deep space—well beyond the GPS constellation. It's also more complicated than just keeping time on the spacecraft, since you need to know how spacecraft time relates to every other instrument that you might use to correlate with it. Converting from the time on the clock on the Kepler satellite to Pacific Daylight Time (with a leap second added part way through) means that you must account for a lot of small, and not-so-small, effects. The arrival of the photons on Kepler can differ by several minutes from the arrival of those same photons in Nevada.

In fact, Pacific time isn't what you really want to use anyway. Initially, the data that were analyzed by the Kepler pipeline and turned over to the science team were the time at the center of the Sun—heliocentric time. But that time adds a four-second variation in the time series because the Sun moves in response to the gravitational attraction of the orbiting planets. You want a reference frame that isn't affected by the planets, or by Kepler itself. We needed the data to reflect the time at the barycenter of the solar system (the barycenter being the balance point among the Sun and all of the planets), so that the variations that we saw in the orbital periods of the planets were from their dynamical interactions with each other and their host star and not our interactions with the solar-system planets and the Sun. Barycentric time is the one reference that removes the effects of all of the motion in the solar system. After some complaints by the planetary dynamicists, all of the times from the spacecraft were converted to Barycentric Julian Date, or BJD. The Julian date is the number of days that have elapsed since noon on Monday, January 1, 4713 BC. (Yes, astronomers can have some strange conventions.)

In January 2010, after nearly thirteen thousand of these high-quality, 50-millisecond-accuracy observations (12,935 to be precise), the pair of CCD chips in module 3 went dark. Nearly five percent of the light-sensitive surface of the Kepler camera stopped functioning completely.

Fortunately, since Kepler rotated once each quarter, the stars that landed on module 3 would still be observed throughout most of the year. But there was now a large blind spot in Kepler's camera, and that blind spot shifted around the field of view with each quarterly roll of the spacecraft. The loss of data from that module actually affected almost one in five targets, with several of the most interesting systems among them.

For example, this included KOI-191 (Kepler-487), one of the five original multitransiting systems with the not-quite hot-Jupiter planet surrounded by a few smaller ones. Another affected system was Kepler-80, the small star surrounded by a swarm of short-period planets, with most of them in a chain of mean-motion resonances. (We saw Kepler-487 in chapter 4 and Kepler-80 in chapter 5.) A third intriguing system that landed on module 3 is KOI-730 (Kepler-223). We also saw this system in chapter 5—the one that was originally mistaken for having a pair of co-orbital planets. With all three of these systems, and the hundreds more that were affected, the data from the lost module would have been helpful in providing additional details into their internal dynamics.

These drawbacks notwithstanding, the loss of module 3 was not fatal to the mission since it just meant less data on some targets. This failure was also not the only reason that data would be lost for given targets from one quarter to another. Just the way the CCD chips were arranged on the camera meant that some targets would occasionally fall in the gaps between them. A notorious example for the science team was KOI-94 (Kepler-89). It landed on the edge of a chip and was only observed during half the year (six months on, six months off).

That four-planet system was the first multiplanet system where we learned that the planetary orbits were aligned with the spin of the star. That measurement was done by examining the Rossiter–McLaughlin effect during the planetary transits. It was also the first system to show two planets that occulted (or blocked) each other. The two planets transited at roughly the same time, but then they both passed simultaneously across Kepler's line of sight. When this happened, the outer planet blocked the inner planet rather than light from the star, which meant there was a slight brightening of the star during the *syzygy*.

(That million-dollar word—useful to pull out when occasion allows—is when three or more celestial bodies lie on a single line, like during a solar or lunar eclipse or when two distant exoplanets simultaneously pass in front of each other and the star that they orbit.)

Following module 3's demise, the spacecraft remained in good health for two largely uneventful years, that is, up until the beginning of March 2012 when the Sun decided to belch out an enormous burst of material. The Sun's magnetic field has a strong effect on charged particles near its surface. Often, the magnetic field will form loops that arc from one location on the solar surface to another. Where these loops penetrate the surface, the gas cools slightly, creating the dark sunspots. The hot plasma near the surface will stream along the magnetic field between sunspots, creating huge prominences of spiraling gas—sometimes large enough to fit all of the planets in the solar system through the arch without touching the sides.

Occasionally, these arcs can pinch off, with the particles now trapped in a large magnetic loop. The loops are buoyant and quickly lift away from the surface and spew the material into the solar system. The resulting coronal mass ejection (or CME) can be pretty spectacular, or devastating, depending upon how it affects you. In most cases, you just get a brightening of the auroras on the Earth as the ejected particles make their way here and become trapped in Earth's magnetic field [20].

The year 2012 was eventful for CMEs. Kepler was hit by a series of five CMEs: two in January, two in March, and one in June. In all cases the impact of the flood of radiation caused the camera to fill with tracks that look like cosmic ray collisions. Kepler data was unusable for some time—between several hours and a couple of days for each one. The March 7, 2012 event was the worst of the set, degrading the data from Kepler to the point that they weren't usable for four and a half days. In July 2012, there was a particularly large CME that burst from the Sun. Had it happened a week earlier, the Earth would have been directly in its path, and the burst of energy could have knocked out communication and GPS satellites and caused some problems for our highly interconnected infrastructure. Fortunately, the Earth missed that one—as did the Kepler spacecraft. Nevertheless, after the series of events in 2012, some of the pixels had a sizable change to their sensitivity, and the overall noise in the detector rose somewhat.

While the Kepler satellite was dealing with these bouts of space weather, the mission itself was being reviewed by NASA for continued funding. Nominally, Kepler was a three-and-a-half-year mission, but NASA has a soft spot for things that work. If a satellite is still functioning, and the data it returns are still scientifically useful, NASA will likely extend the mission—adding more funds to the pot to keep it running. The decision to extend missions comes through senior reviews, where the productivities of different NASA missions are compared and a pool of money is divided among them. The 2012 senior review examined nine NASA missions, deciding which ones to keep running and which to shut down. This review was a big deal for us, especially since Kepler was competing with its older, and often more famous, siblings. The candidates in this round included such notable crafts as the Hubble Space Telescope, the Chandra Space Telescope (the X-ray counterpart to Hubble), the Spitzer Space Telescope (the infrared counterpart to Hubble), and both the Fermi and Planck satellites (major observatories for work in cosmology).

Here was Kepler, going head-to-head with three of the four *Great Observatories* (Hubble, Chandra, and Spitzer), and leading missions to study high-energy astrophysics and the cosmic microwave background. We needed to put together a strong case for continued funding. At the same time, we still needed to conduct the scientific investigations that Kepler was built to make. Our preparations for the senior review were happening throughout 2011—the same year that we announced so many of our most consequential discoveries. The mission managers wanted to task someone with overseeing our request for an extension. I know that at least two people were interviewed to fill the spot—myself, and the one who got the job, Steve Howell.

Steve had been a long-time member of the ground-based Kepler Follow-up Observing Program—conducting many of the important *speckle* observations to identify background stars that could contaminate the exoplanet detections. He was brought to NASA Ames, and worked throughout the year to make our case to the powers that be at NASA. Eventually, after four days of deliberation among the review panel, Kepler was listed among the missions given the green light to continue. We fared quite well, in fact, ranking third—after Hubble and Chandra. With that result, we were approved for an additional two years of funding—two more years of drinking from the proverbial fire hose.

Unfortunately, it was also around this time (mid-2012) that the science office noticed one of the reaction wheels was having problems. These wheels are essential to maintaining the precise pointing of the telescope. If the telescope begins drifting, one or a few of the wheels will start to spin. Since every action has an equal and opposite reaction, the spinning reaction wheel causes the telescope to rotate in the opposite direction. Given the relative sizes of the wheels and the spacecraft, it takes a lot of reaction wheel spin to produce a small response in Kepler—which is why they are so good at making small adjustments. Three wheels are needed to keep things running, and Kepler had only one spare.

Reaction wheels are designed to operate smoothly for years with speeds of several thousand rotations per minute (rpm). The standard for the reaction wheels on Kepler was 3000 rpm. The specification for the mission was that they needed to last for three and a half years. Then, if all went well, the mission would be extended for additional time—presumably until the hydrazine ran out. Given the surprising longevity of some missions, it is easy to assume that they will always live to a ripe old age. However, this doesn't always happen. A consequence of setting relatively low expectations up front is that moving parts often fail shortly after the nominal mission completes.

After three years of use, reaction wheel 2 showed signs of wear. When it spun there was more friction than expected. Though it could still operate, this friction was a concern. Once a wheel starts to go, the additional stress on its internal workings can cause it to wear faster in a vicious downward spiral. On July 14, 2012, wheel 2 failed. After a few tries at bringing it back online, with hopes that it may be salvaged, the science office gave up and turned it off. It took six days to bring the spacecraft back online with the remaining three wheels. With wheel 2 gone, there was no recourse if a second wheel failed since there had only been one spare. Unfortunately, the spare (wheel 4) had already begun acting up. In what is probably the only upside to the failure of wheel 2, we now had some diagnostic information about how it failed. We could use this information to monitor the health of the other wheels—looking for telltale signs of impending demise.

The original three-and-a-half-year mission was complete, the team had just passed a review by senior NASA personnel, and we were six months into the extended mission. Hopes were high that we would still see several years of successful science operations. There was a lot still to learn about the planetary systems that Kepler had discovered, and we were just entering the realm where we could find what Kepler was originally designed to see—Earth-sized planets in Earth-like orbits around Sun-like stars. But six months after the failure of wheel 2, the issues with wheel 4 couldn't be ignored.

Not only did the degradation of the wheel affect the stability of the pointing of the spacecraft, but the friction increased the temperature of the fixture that it was housed within. That increased temperature affected both the focus and the photometric precision of the telescope. Space missions are, if nothing else, particularly sensitive thermometers, as the electronic and mechanical systems respond to changes in temperature. In January 2013, the engineers tried bringing the wheel to rest in the hopes that whatever was causing the issue would subside. Perhaps it just needed the lubricant to be spread more uniformly around the wheel. Well, when they started the wheel back up, it was worse. The spacecraft could still function, but it looked like it was running on borrowed time.

Four months later, the mission manager reported that

the spacecraft entered thruster-controlled safe mode at about 7:30 p.m. PDT on Wednesday, May 1, 2013. The recovery operation began at about 5 p.m. PDT on Friday May 3, 2013, after engineers had verified that the spacecraft was otherwise operating normally. The spacecraft responded well to commands and transitioned from thruster control to reaction wheel control as planned.

Despite this positive result, ten days later Kepler was once again in safe mode. Now, the mission manager reported that

> we attempted to return to reaction wheel control. . . . Initially, it appeared that all three wheels responded and that rotation had been successfully stopped, but reaction wheel 4 remained at full torque while the spin rate dropped to zero. This is a clear indication that there has been an internal failure within the reaction wheel, likely a structural failure of the wheel bearing.

With this second wheel now out of commission, the spacecraft could not maintain the necessary pointing stability. Despite numerous attempts to try to revive it, the science operations for the Kepler mission ended.

Emails filled with condolences circulated among science team members. The mission was producing such good data, making such monumental discoveries. That a pair of small wheels would cripple it was an unbelievable shock. There was a pit in my stomach at the loss of such an instrument. Half of the hydrazine was still on board, and if the wheels kept working, the mission would have been only halfway through what it was capable of doing. The extended mission was underway and we were all geared up to work on Kepler data for the rest of the decade. (I had just completed a number of job interviews at different universities, with Kepler and its extended mission as a central aspect of my planned research program.)

Six months after its warranty expired, the Kepler mission was over, and news of its demise spread through the community. I'm sure many of my thoughts matched those of my colleagues. "A half-billion-dollar mission hinged on such a small part. How much more would it have cost to have two spare reaction wheels instead of just one?" But these thoughts couldn't change the past, and over the next few weeks (while recovery options were systematically eliminated) we all came to terms with the loss.

We had a decision to make regarding what to do with the spacecraft. The camera still worked. Two of the reaction wheels still worked. And it still had fuel. We could operate the spacecraft under these conditions.

Even with the loss of precision from the lack of fine-pointing capabilities, the Kepler photometer was still the best that the world had to offer for these kinds of observations. NASA invited the Kepler team members to put together a proposal with what to do with the satellite—and to compete it in the 2014 senior review. This would be another review like the one that took place in 2012, which would again pit Kepler against the icons of the NASA fleet.

By now it was near the end of summer 2013, and the senior review was to take place in six months' time. A call went out from the Kepler Science Office to the entire astrophysics community for suggestions about what to do with the crippled, but still formidable, telescope. It basically read, "If we were to give you control of a slightly defective six-hundred-million-dollar telescope, what would you do with it? Oh, we need your proposals, in writing, in one month. Oh, and you're competing for money against Hubble, and Spitzer, and Chandra, and Fermi, and. . . ." You get the point—the stakes were high and the time was short [72].

Several of the Kepler science working groups started making their cases for where Kepler should point and for how long. Two proposals came from among the dynamicists in our group. We recommended that Kepler keep looking at its original field of view, despite the worse precision we would get from the new data. The telescope drift could be compensated for by periodic thruster firings. One of the proposals outlined the science case from the standpoint of the long-period, eclipsing binary stars that we would detect with the longer observational baseline. I was part of the other proposal, which we called Kepler-II, but soon shortened to K2 in our verbal discussions (the first person I heard use the term was Dan Fabrycky, a long-time member of our group who had just started as faculty at the University of Chicago).

Like our eclipsing-binary counterparts, our recommendation was to keep pointing Kepler toward the same part of the sky. There were a number of science opportunities for exoplanets with this approach. Despite the more limited data quality, we could continue to make transit timing measurements for a significant number of systems which showed large dynamical interactions among the planets. Such observations would produce improved planetary mass measurements for some of the smallest planets yet detected. Another advantage that we outlined was the ability to measure the orbital periods of long-period planet candidates. There were a number of planets that appeared to circle their stars only once every three or more years, and additional data from the original Kepler field could provide rare information about planets beyond the orbit of Mars. This approach could also pin down ambiguous orbital periods for a few candidates, as there were a handful of interesting planets where a gap in the data occurred exactly halfway between two transit measurements. It wasn't clear with these planets whether the measured orbital periods were correct, or too long by a factor of two.

The downside of our proposal looking at the Kepler field of view was fuel consumption. Keeping the telescope pointed in that direction would require a lot of hydrazine. Part of the reason for the needed thruster firing is that the Sun's light, bouncing off the telescope and solar panels, exerts a pressure on the spacecraft causing its orientation to constantly drift. Normally this would be corrected through the reaction wheels. With two wheels you can adjust the spacecraft to keep it pointed along a line, but you can't prevent it from rotating about that line. So the telescope could be pointed in the right direction, but the camera would circle about that axis—with the target stars shifting from one module to another. The third wheel is essential for fixing its orientation in all three directions, and using the hydrazine thrusters was the only way to compensate for the missing third wheel. These thrusters have a minimum duration that they will fire and can't make corrections that are too small, so relying on them meant that the pointing drifts would be pretty severe.

Ultimately, forty-five mission concepts came from the astronomy community into the science office. There was clearly a lot of valuable science that could be accomplished, even with the reduction in photometric precision. While these proposals were flooding in, an engineer at Ball Aerospace, Doug Wiemer, proposed a new method to help keep the spacecraft stable. The solar panels that surround the Kepler satellite are arranged symmetrically around its hull. If the sunlight struck directly on the ridge at the center of the array, the solar radiation pressure could help prevent the spacecraft from rolling. It would still be a precarious balancing act—as long as the light was striking the ridge directly, it would be stable, but if it started to roll, the radiation pressure would be unbalanced and would cause it to roll even faster. Nevertheless, keeping the sunlight shining on the apex of the array would provide better stability than other options.

The downside of the approach outlined by Doug was that it couldn't point just anywhere in the sky, the satellite had to point within the plane of its orbit. And it couldn't point in the same direction all the time. For one thing, it needed to keep the solar panels pointed toward the Sun. The balancing act could last for about eighty days at a stretch before the satellite needed to shift to a new field of view. The upside was that a lot of the science outlined in the different proposals could still be done with this restriction on where Kepler could look: it would be done with better photometric precision because the view would be more stable than otherwise anticipated, and it wouldn't require as much fuel to operate—extending the lifetime of the mission.

While the Kepler Science Office was putting together the case for this new mission concept, the team at Ball Aerospace got to work practicing the new observing method. They made weekly tests of different strategies over two and a half months in the fall of 2013, making incremental improvements to the operating procedures. One improvement, for example, allowed them to make smaller corrections to a spacecraft roll by simultaneously firing two opposing thrusters—ones that would try to spin the telescope in opposite directions. The forces from the two thrusters mostly canceled each other, and the small difference (less than ten percent) could keep the spacecraft oriented to within four-thousandths of a degree, yielding even better photometric precision than originally anticipated [16].

The engineers were also able to get the thruster corrections down to incredibly short periods of time—making it so that less data would be lost during attitude corrections. Every six hours there would be small corrections that took less than a single science cadence to complete— so that no data would be lost. Then, every two days there would be a larger correction that would take slightly more than a half hour. This approach meant that only a single data point would be missing every two days. As a bit of icing on this promising cake, one of the week-long tests in December produced a planetary transit, confirming the promise of this approach. Operating in this mode, Kepler could even continue its quest to find small planets.

Here, it is worth noting that a NASA *mission* is typically the combination of an instrument (the flight segment) and the science operations that go along with that instrument (the ground segment). If you change either of these components, you generally have a new mission. It is common for other satellites to be repurposed as new missions. For example, there is the Deep Impact mission, where a large copper slab was slammed into a comet while the satellite filmed the results from a distance to see what came out. Deep Impact was similar, in some ways, to the recent Double Asteroid Redirection Test (DART) mission that was designed to change the orbit of the moon of an asteroid by slamming into it. After Deep Impact finished its operations using the projectile, and the primary mission ended, the satellite still had a working telescope and camera. So NASA repurposed it as EPOXI, a mission with two science programs. These were the Deep Impact eXtended Investigation (DIXI), where it went hunting for another comet to observe, and the Extrasolar Planet Observation and Characterization (EPOCh) mission, where it made observations of transiting extrasolar planets as it traveled to its new cometary target.

Usually, developing a new mission from a repurposed spacecraft takes a year or more, beginning with community input and running through flight testing. For Kepler, this process was given only a few months. The proposal went to the NASA senior review panel in early 2014. This time it came out on top—the highest-ranked proposal, beating out a stacked field of contenders. In this review, even the Spitzer Space Telescope was slated for decommissioning. The Kepler mission was over, but it would be reconstituted in the form of a new mission, K2. We were all feeling fortunate with K2's success. On May 16, 2014 NASA approved the K2 mission, and observations in its first *campaign* began two weeks later on May 30.

As long as it operated, K2 would run a series of campaigns where it pointed to a field in the ecliptic plane (along the constellations of the zodiac) and stared at it for two and a half months. Once finished, it would point to a new field in the ecliptic plane and the process would repeat. Over the next few years, K2 campaigns would cover a sizable fraction of the sky along the plane of its orbit. This observation footprint dovetailed nicely with a successor mission to Kepler called the Transiting Exoplanet Survey Satellite (TESS), which we will discuss in the next chapter. TESS would scan most of the sky, except for the ecliptic, where K2 was searching.

Initially, the science team thought that the photometric precision would degrade by more than a factor of ten, down to a few hundred parts per million (instead of the twenty parts per million for the original mission). As improvements in spacecraft handling and improvements in the analysis software made their way into the operations, the photometric precision continued to improve, eventually reaching forty-four parts per million—a decline of only a factor of two. Though the weakened precision and the shorter campaigns meant that Earth-sized, habitable-zone planets were off the table, there was still a lot of discovery space that K2 could explore.

From May 2013, when the second reaction wheel failed, and May 2014 when the K2 mission started science operations, the science team, or rather the agenda of the science team, split in two. With K2 running, new data were coming back to Earth. Since the observing field was changing every few months, we needed to identify new targets, downlink and ingest those data into online repositories, and conduct new ground-based observations to sift through the K2 results. At the same time, the original Kepler data were still hot off the presses and there remained thousands of exoplanet candidates to sort through, lots of tests to run, and the most important scientific objectives of the mission to address. This work on the original Kepler data, and the studies that result from them, are still going on a decade later.

Among the members of the science team, several turned their attention entirely to the K2 mission, several had a foot in both camps—working on whatever problem needed to be addressed at the time—and several worked almost exclusively on data from the original Kepler mission. I was part of this third subset. This separation of the focus of different science team members was driven primarily by individual interests. The group that I worked with was investigating questions that were best studied in the original data. We needed the long-term, continuous observations, and the uniformly observed and calibrated data that the original mission provided. Most of this book has focused on the original mission. But there were a lot of groundbreaking discoveries that came out of K2, which have weighty implications for our understanding of planets and their histories.

Some of the most impactful exoplanet systems were discovered, or characterized, by the K2 mission. My personal favorite is WASP-47. This is the hot-Jupiter system that has small, nearby planets—the primary counterexample for the prevailing scenario where hot Jupiters are lonely. The hot Jupiter was discovered in 2012 with the Wide Angle Search for Planets (WASP)—a ground-based survey looking for planetary transits. Three years later, Doppler measurements of the system confirmed the existence of another Jupiter-mass planet with a high eccentricity, on a one-and-a-half-year orbit. Hot Jupiters are often found with distant, eccentric, outer companions. So this discovery was not unprecedented, but it was interesting—in part because the outer companion's eccentric orbit causes it to pass through the habitable zone of the host star.

What was unprecedented for WASP-47 was the 2015 K2 observations that showed transits for two additional planets. One was a hot Earth on a one-day orbit, a member of that population of similar planets that are dynamically isolated from each other. The second was another small planet just outside the orbit of the hot Jupiter, with an orbital period just over nine days. Being the first system known to have a hot Jupiter and neighboring small planets was a major contribution to exoplanet science from K2. Exceptions to the rules contain a disproportionate amount of information about what is possible in the universe. And the K2 discovery of these small planets in WASP-47 made a large impact on our understanding of the formation of hot Jupiters [73, 74].

For me, WASP-47 was a unique mixture of stuff that I had worked on over the decade before its discovery. Beginning in graduate school, I studied transit timing variations—WASP-47 had those. I had done several searches for small planets near hot Jupiters—WASP-47 had those. I worked to characterize the population of hot Earths on one-day orbits—WASP-47 had one. I worked on planet pairs with orbital period ratios near 2.2 (just wide of the 2:1 resonance), and WASP-47 had one of those too. The alignment of this system with my research history was uncanny. In the unlikely event that there is ever a crewed mission to that system, I hope that they use my super-duper-efficient method of boarding passengers when they embark.[1]

Another system where K2 played a pivotal role is the TRAPPIST-1 system. The TRAPPIST survey (which stands for TRAnsiting Planets and Planetesimals Small Telescope) uses a ground-based, robotic telescope sixty centimeters (two feet) in diameter. While the telescope is located in Chile, the project is headquartered in Liège, Belgium.[2] The star in TRAPPIST-1 is a tiny, M-dwarf star less than ten percent of the mass of the Sun. If it was much smaller than that, it wouldn't be a star at all, since at about eight percent of the mass of the Sun, a wannabe star doesn't generate the core temperatures necessary to power nuclear fusion. In May 2016, the TRAPPIST survey found three planets orbiting TRAPPIST-1. Later, observations from the Spitzer, Hubble, and K2 space telescopes added four more planets to the list. This brought the total number of planets to seven, with as many as four (yes, four) planets located within the habitable zone of the host star [50].

This star is so small (about the size of Jupiter) and so cool (less than half the temperature of the Sun at about 2500 kelvin, or 4000°F) that the habitable zone is just a stone's throw from the stellar surface. The TRAPPIST planets orbit with periods of 1.5, 2.5, 4, 6, 9, 12, and 19

---

1. In 2008, right before my work on the Kepler mission started, I wrote a paper that outlined the optimum method to board airplane passengers.

2. Liège is the town the Germans sieged in World War I using the 420-millimeter "Big Bertha." It is also home to the best variety of Belgian waffles. Brussels-style waffles are like typical waffles dipped in chocolate or dusted with powdered sugar. Liège-style waffles are made with a cookie-dough-like base filled with giant sugar crystals that caramelize in the waffle iron.

days. The 4, 6, 9, and 12-day planets are spread from the inner edge to the outer edge of the habitable zone. The orbital period ratios of the TRAPPIST-1 planets, which are the key observable for under-standing their orbital histories, fall near ratios of 8:5, 5:3, 3:2, 3:2, 4:3, and 3:2—indicating that their past included converging migration and capture into mean-motion resonance through interactions with the planet-forming disk.

The planets orbiting the star with these period ratios create large transit timing variations. These variations, in turn, allowed scientists to make precise measurements of their masses. These measurements, combined with the radius measurements from the transit observations, produced some of the best measured planetary densities among the five thousand or so exoplanets in our databases. One conclusion from these measurements is that most of the TRAPPIST planets have little or no atmosphere. They are rocky like the Earth, except that they likely have a fair amount of water—typically a few percent of their mass, but per-haps as much as a third of their mass if their internal structure differs significantly from Earth's.

Because they orbit so close to their host star, these planets will often have their orbital and spin rates coupled—keeping the same face pointed toward the star at all times. However, this state can change peri-odically for individual planets if their orbital periods vary too quickly from the mutual gravitational perturbations of the planets. One of my graduate students, Cody Shakespeare, looked at how the orbital period variations would cause the spin periods to decouple from, or recouple to, the orbital periods from time to time. Those changes, in turn, would affect the stellar illumination of the surface and ultimately the planetary climates.

The fact that there were so many planets orbiting a distant star, and the fact that several of those planets (likely with some water) were in the habitable zone of that star, generated no small excitement. TRAPPIST could be a rare *multihabitable* system, where multiple planets have the necessities of life. If that were true, then all sorts of interesting scenarios would be possible. We have, on the Earth, meteorites that were chipped off the surface of Mars, presumably from an ancient collision that Mars had with an asteroid or comet. (We can tell that they come from Mars because the compositions of the meteorites match the composition of Mars and not the composition of other known rocky material in the inner solar system.) After wandering through the inner solar system for some time, probably tens of thousands or millions of years, these fragments found their way to the Earth.

The fact that there is a natural mechanism to transfer material from one planet to another in the solar system has led some scientists to speculate that life on Earth could have originated on Mars first (because it cooled faster, and therefore had liquid water earlier than the Earth). After an impact on Mars, life-bearing fragments could have transferred to the Earth—spawning life on our planet. Panspermia, the idea that life spreads from one planet to another, is an old idea. We already see it on a smaller scale on the Earth. Life from one side of a river can often make it to the other. Or life from one island can make its way to another island. Going beyond the migration of life across a continent or ocean, the next largest scale for panspermia would be having life travel back and forth between a planet and a moon, or between two habitable moons orbiting the same planet. One more step in size and you arrive at interplanetary distances, such as between the Earth and Mars.

I don't necessarily buy this hypothesis for life's origins on Earth, but it is plausible given what we know about the dynamics of the inner solar system and the resilience of some forms of life under extreme conditions. (Tardigrades, or "water bears," seem to be nearly indestructible, like microscopic cockroaches, or Terminator robots. They have survived, and reproduced, after spending ten days in the vacuum of space.) In a multihabitable system, especially with planets that are significantly closer together than the Earth and Mars, microbial life may have multiple opportunities to spread from one planet to another in the system. I wrote a paper hypothesizing about this possibility just prior to the discovery of the TRAPPIST system. My focus was on Kepler-36, with its two neighboring planets, but TRAPPIST would have been even more interesting.

Extrapolating panspermia to the other extreme, the opposite of life simply crossing the road would be life from one star system drifting to another—eventually populating the galaxy. This scenario might seem far-fetched, but in 2017, Robert Weryk examined images from the Pan-STARRS telescope in Hawaii and found a piece of debris from a distant star passing through the solar system. We could tell by the shape of its orbit that it wasn't from the neighborhood. Oumuamua, as it is called, was the first piece of interstellar material ever detected. The fact that it was seen only a few years after we were capable of finding it indicates that such material may frequently pass through the solar system.

Without needing to consider such large distances, one could also imagine life traveling between planets orbiting two different stars in a multistar system. In a particularly lucky system, one might find intelligent life developing independently on multiple planets (perhaps following their divergence from the common ancestor that made the interplanetary trip on a rock fragment). In a system like that, different inhabitants could readily devise ways of communicating with each other centuries, or millennia, before they develop the necessary technology to travel to each other's homes.

This hypothetical scenario implies that they developed simultaneously. But the reality, given how much time evolution requires, and how briefly humans have been on Earth, is that it would be far more likely for one planet to have intelligent life long before the other. Moreover, given the random nature of asteroid and comet collisions, you could imagine having one planet keep their equivalent of the dinosaurs while another moves on, simply because one gets hit and the other one doesn't. Anyway, the point of all this is to show that TRAPPIST is an incredibly interesting system that opens the doors to a rich collection of future investigations.

Since its discovery, TRAPPIST has been a playground for scientists to compare with their models. Its planets have among the most precisely measured masses and sizes, so we can put interesting constraints on their bulk composition and structure. For example, a recent study showed that a single composition (meaning relative abundances of different planet-forming materials such as metals and rocks) can reasonably describe all of the planets in the system to within the uncertainties of the planet measurements. This does not mean that the planets all have the same composition, but it does imply that their compositions are similar. We know that the solar-system planets cannot be described by a single planet composition, and a sizable fraction of the multiplanet systems from Kepler also appear to have non-uniform composition.

Beyond discovering new planetary systems, K2 also made discoveries in other areas of astronomy. Since each field of view is near the ecliptic plane, K2 had occasion to observe objects in the solar system. These included the planets, their moons, and small bodies like asteroids and comets. There were observations of Trojan asteroids in the orbit of Jupiter, objects beyond Neptune, main-belt asteroids, and Centaurs—small asteroid and comet-like objects with orbits that often cross those of the giant planets, limiting their survival in the solar system to only a few million years before they get ejected from a close encounter.

K2 also surveyed other star clusters, including young star clusters like the Pleiades and the Beehive clusters, and older ones like the cluster M67 where the stars are similar in age and composition to the Sun. It looked for binary stars, planets, pulsations, and rotations. In some campaigns, the spacecraft was oriented in such a way that simultaneous ground- and space-based observations of the targets were possible. K2 measured several supernova explosions, with unprecedented photometric precision. It also gathered data on different kinds of active galactic nuclei—resolving detailed changes in brightness of the jets of radiation that are powered by supermassive black holes. All of these data contributed new insights into the properties of these astronomical objects from the unique combination of continuous observations with high-quality data.

As the K2 mission continued to make new observations and new discoveries, it also continued to age. In January 2014, as the K2 mission was being pitched to the senior review committee, a second pair of CCDs failed. The first pair of chip failures happened early in the Kepler mission with the pair on module 3, centered along one of the edges of the photometer's observing footprint. Here, it was the pair of chips on module 7 that failed, located on one corner of the middle ring of detectors.

Despite these two failures, the rest of the telescope seemed to be functioning well, and K2 continued its series of campaigns. However, in the spring of 2016, just at the start of Campaign 9, the spacecraft was found in *emergency mode*, its lowest operation state. A glitch in an electronic component caused the telescope to have large excursions from its nominal pointing direction. That tripped the onboard "don't stare at the Sun" protocol. This was the first time the spacecraft had ever entered emergency mode, so the engineers brought it back to life cautiously. Entering emergency mode also reset the onboard computer, causing the loss of any onboard science or engineering data. It also caused the onboard clock to go out of synchronization with the rest of the data that it had gathered over the preceding seven years. The clock issue took time to sort out, and required that four seconds be added back to the timestamps on some of the data—a big deal given the tight timing of the observations.

With each new anomaly from the spacecraft, there was always concern that it might be the last. And since we weren't in constant communication with the satellite (it does have a job to do, after all), the anomalies were always discovered after they had already occurred. In late July 2016, the entire camera turned off. A few days later, when the team learned that it was off, they tried turning it back on. Within a couple of days it was again collecting science data. But now the pair of chips in module 4 were questionable—a third malfunctioning module. After diagnosing the state of the camera by downloading the pixels for a few stars that were spread across the different parts of the camera, all of the functioning modules showed what they were expecting to find—except module 4, which never turned back on. A remote investigation as to the cause of these three CCD failures pointed to blown fuses. Plate 4 shows the final image ("last light") taken by Kepler. You can see the gaps from the three failed modules when compared with the first-light image, plate 3.

Despite these hiccups, NASA was determined to keep taking data until there was no choice but to shut down. Funding for continued operations was granted through the end of NASA's fiscal year in late September 2019. By that time, we didn't expect any fuel to remain.

Despite the scares of 2016, with the emergency mode and the failure of module 4, Kepler still had gas in the tank. For nearly two more years (from Campaign 10 in the summer of 2016, through Campaign 18 in the summer of 2018) it took data with little additional drama. As June transitioned to July 2018, however, pressure in the fuel lines began to drop—Kepler was now running on fumes. Roughly one month passed before there was an attempt to make any new observations, and when it did start up, it immediately went into safe mode.

There were still observing plans on the table in late 2018. Campaign 19 was underway (though its performance had been degraded from lack of fuel) and the mission announced its target list for Campaign 20 with nearly 15,000 long-cadence targets with half-hour exposures, and 210 short-cadence ones with one-minute exposures. The selected field included stars from the Pleiades and Hyades star clusters, as well as a number of very young stars (less than ten million years old—the stellar equivalent of a one-month-old infant). The mission noted large numbers of white dwarfs, brown dwarfs, flare stars, solar analogs, and classical variable stars on the list of anticipated targets.

Campaign 20 was scheduled to run from October 15, 2018 through the beginning of January 2019, but on October 1, NASA placed the spacecraft into no-fuel-use sleep mode. It was roused again on October 12 to maneuver into position to download the data from Campaign 19. The data made it back to Earth on the 15th, and Campaign 20 started. Four days later, Kepler was found in no-fuel-use sleep mode again, and on October 30, 2018, NASA announced that the ship would be retired [75, 76].

Nearly four centuries ago, Johannes Kepler passed away on November 15, 1630 at the age of fifty-eight. His work broke ground for our modern understanding of planets and planetary systems, and first predicted transits of Mercury and Venus across the Sun. He was the namesake for NASA's planet-hunting mission for good reason. On November 15, 2018, 388 years after his death, NASA's Kepler mission was deactivated following nine years, seven months, and twenty-three days of operations.

9

Kepler's Legacy

Before Kepler had lifted off the ground, successor missions were already being designed and proposed. Given the long development times for almost anything that travels to space, this advanced planning is expected. At an early Kepler Science Team meeting in Boston, David Charbonneau stood in front of the team and unveiled the rudiments of what became the NASA TESS mission (Transiting Exoplanet Survey Satellite). This mission would follow the Kepler results with a different survey strategy that would cover more sky, but with a trade-off of less depth and for shorter duration [77]. While this was happening in the United States, the Europeans were already developing the PLATO mission (PLAnetary Transits and Oscillations of stars) [78]. This mission was a successor to both Kepler and the European CoRoT mission, which had preceded Kepler by a few years. All of these missions, when you consider what they would find, compete with Kepler in some ways, and complement it in others.

One of the challenges we face when working with Kepler data is the dimness of the target stars. Because Kepler was designed to find small transiting planets in one-year orbits, it needed to stare at its targets continuously for several years. The continuous observations were an essential aspect of its design. Earth, if viewed from a distance, would transit the Sun in only six hours. So, if Kepler were to miss the wrong six-hour interval in its four years of observations, it could render most of the observations on a given target nearly useless. That need to keep the spacecraft pointed in the same direction throughout the mission meant that all of the targets it was to observe needed to lie along a single line of sight. It couldn't survey the whole sky to find planets.

Because all the targets were along a single line they tended to be far away, since there aren't enough nearby stars along that line of sight to yield a good sample of potential planetary systems. The more distant the targets are, the dimmer they appear—which, in turn, makes them harder to observe with any instrument other than Kepler. Consequently, if you wanted to study a Kepler planet, beyond simply finding that it exists, you either need to find one around a rare, bright target, or you need access to very large telescopes. There are only a handful of large telescopes, so there is a lot of demand on their time, and even with those telescopes it is both costly and time consuming to make much progress on dim systems. It's no surprise that relatively few of the Kepler discoveries have since been studied in detail.

The TESS mission was designed to solve the issue of distant, dim targets. TESS would survey the whole sky, rather than a single line of sight, so that there would be more bright targets to observe. Compared with Kepler, the TESS targets are ten times closer to the solar system and a hundred times brighter. Being a hundred times brighter means that ground-based observations reach the same sensitivity in one-tenth the amount of time—minutes instead of hours. That makes a big difference if the signal that you are looking for, like measuring the composition of a planetary atmosphere by observing the light that passes through it, only occurs during a planetary transit—which, right off the bat, limits your observation to only a few hours.

In order to survey the entire sky (or ninety percent of it), you can't stare at everything without interruptions the way that Kepler did. TESS only looks in a particular part of the sky for a month before it changes direction and looks somewhere else. Each month, it has to start over with a new field of view and new target stars. Continuous long-baseline observations of a single system is challenging at best. You might catch a small planet on a one-year orbit during that month, but it would only be a single transit. You wouldn't be able to determine the orbital period, and it would take more follow-up observations to demonstrate that it wasn't a spurious signal.

The TESS scientists did find a way to squeeze in longer(ish)-baseline observations for some targets. The TESS telescope is an array of four large digital cameras (these cameras use lenses instead of mirrors— they look a lot like the cameras you see on the sidelines of sporting events). Unlike the Kepler field of view, which was essentially a big square, the TESS field of view is a large four-by-one rectangle—each of the four cameras sampling a square of the sky, 24 degrees by 24 degrees (compared with Kepler's 10-by-10-degree field of view). The team centered the first square of the camera array (camera 4) on a single part of the sky. When the time came to observe the next field, the camera array rotated around the center point of that camera's field. Thus, most of the sky seen by that camera remained in the field, and observations of those stars continued uninterrupted. The resulting survey footprint was a series of thirteen partially overlapping, flower-like petals that cover nearly half of the sky. After one year of observations, mostly in the southern hemisphere, the spacecraft flips over and surveys the half of the sky that is primarily in the northern hemisphere. Plate 10 shows the Kepler, K2, and TESS fields of view on the night sky.

The portion of the sky that was continuously observed is the ecliptic pole—directly above the axis about which the Earth orbits. Each petal extends down, nearly to the plane of the Earth's orbit. (There is about a 12-degree swath near the ecliptic—6 degrees each to the north and south—that isn't observed.) The choice to have one camera centered on the ecliptic pole wasn't arbitrary. In fact, it was recommended by the people that NASA chose to review the proposal before it was selected for flight. That portion of the sky is also where the James Webb Space Telescope is able to view continuously. Thus, not only did TESS observe for a full year the targets that are found in this region, but they are also accessible to large amounts of follow-up observations with the most ambitious telescope that NASA has launched to date, JWST.

Most TESS discoveries, however, will be planets with orbital periods less than one month. These are unlikely to be habitable, except for perhaps those that orbit the smallest stars. And a large fraction, if not an outright majority, of planets that exist in the TESS systems will not actually be seen in the TESS data—they will be found through follow-up observations. These limitations to the TESS program are not fatal to the mission. Even with the Kepler data, we need other observations to fully characterize the planetary systems, like Doppler measurements or follow-up transit data from other instruments. Kepler, save for lucky exceptions, was only able to see planets that had orbits less than one year. So analogs of the solar-system giant planets, Jupiter, Saturn, and especially Uranus and Neptune, are almost entirely outside Kepler's sensitivity—despite their clear importance in the solar system. (Earth survives due to their benevolence, since relatively small perturbations to the orbits of the giant planets can completely destabilize the orbits of the rocky planets—as we saw with Mercury and Jupiter.)

TESS also has a slightly different observing cadence from what Kepler used. Kepler observed most targets for thirty minutes at a time, with 512 targets being observed every minute. In addition, Kepler only stored the relevant pixels being sampled. All TESS target stars, on the other hand, are observed every two minutes, and the entire camera is sampled every half hour—not just the "postage stamps" of Kepler. So the time resolution of TESS is superior to Kepler in many regards.

Comparing the advantages and disadvantages of both missions against each other, one may wonder why we didn't build TESS first. Its targets are brighter, it covers more sky, more often, and it's half the price. At this time, TESS has a longer planet candidate list than Kepler. It might seem to have been a wiser choice to start there, and then move to the more expensive Kepler mission if it was needed. But, at the time that Kepler was conceived, and even by the time Kepler was selected, no one would have expected that planetary systems filled with super-Earths and sub-Neptunes on ten-day orbits would be so common. If, instead of having half of the stars in the sky filled with these easy-to-detect, short-period planets, most stars had planetary systems like the solar system, TESS may not have found any planets at all. It is easy to imagine that had TESS been proposed in the 1980s and 1990s as a cheaper alternative to Kepler, before the detection of hot Jupiters and any compact multiplanet systems, it would have failed to receive any funding because it would have expected to find no planets.

Kepler's design was centered on finding another Earth, along with whatever else happened to be there—which turned out to be quite a lot. Its discoveries showed that TESS would be successful, and they play an important role in interpreting TESS's findings. In general, the TESS sensitivity to planets declines for orbital periods beyond one month. But, for most of those cases, we know from the Kepler data what kinds of planets likely live in those systems out to orbital periods of about a year (where Kepler's sensitivity starts to decline). TESS discoveries, so far, have been more of the same Kepler fare, but brighter and easier to study in the long run. We know from Kepler that there will be a lot more planets to find in the TESS systems if we spend time observing them. Kepler didn't have that luxury.

In the European scene, the successor mission to Kepler is the PLATO satellite. It was proposed in 2007, just prior to Kepler's launch. It was selected as one of four finalists for a competition among medium-class missions by the European Space Agency, ESA. The astonishing success of the Kepler mission in the early 2010s provided some reassurance about the benefit of getting another telescope in space to look for planets. In 2014, ESA announced PLATO as the winner, with a planned launch a decade later. That timeline has since slipped, but that is largely because the timelines of all space missions slip.

PLATO is like a mixture of Kepler and TESS, in terms of both its observing strategy and the nature of the instrument itself. The business end of the satellite is an array of refracting cameras much like TESS. But there are twenty-six of them instead of four, making the collecting area more similar to Kepler (several times larger, in fact). PLATO will park itself at the L2 Lagrange point, where the James Webb Space Telescope is located. For PLATO, the fields of view of the individual cameras overlap such that, in the center, where the largest number of cameras converge, the sensitivity compares favorably with Kepler. Meanwhile, the periphery has fewer overlapping cameras and will have a reduced sensitivity, more like TESS.

While Kepler's observing strategy was to stare at a single part of the sky continuously for four years, and the TESS observing strategy is to work its way across the whole sky one month at a time, PLATO plans to do some of both. It will have a long stare in the northern hemisphere (about two years in duration), a long stare in the southern hemisphere, and several fields that it observes for a few months. At least, that is the plan right now. We'll see what the final decision is on this front. There are compelling reasons, for example, to point at one part of the sky for a very long time, longer than the duration of the Kepler mission, in order to characterize planetary systems beyond the one-year orbital period. There are also compelling reasons to return periodically to just the Kepler field of view to monitor the known planets that are there. For now, the plan is a few long stares and a handful of short ones. Most of the areas of the sky that PLATO will observe will have been seen by TESS, but now there will be more, and higher-quality, data. Some of the expected observation fields also cover the original footprint of the ESA CoRoT mission.

One of the long stares will include the Kepler field of view, providing valuable new information on the planetary systems that Kepler saw. By revisiting the Kepler field, PLATO will extend the total duration of the observations from four years to six or more. Those data will be spread out over two decades. Such observations are useful for monitoring the ongoing dynamical interactions among the planets in the Kepler systems—allowing improvements to mass measurements from transit timing variations, refinements to various orbital parameters for the planets, and likely detecting or confirming longer-period planets in the Kepler systems. When it comes to dynamical systems like planets orbiting stars, there is no substitute for long-baseline time-domain observations—"longitudinal studies" if you will.

Beyond exoplanet science, PLATO will also produce data to do the same stellar astrophysics that Kepler provided, but a decade or two later. These observations will allow us to see changes in stellar activity over the course of many years. We know, for example, how the Sun goes through an eleven-year sunspot cycle. Dark spots on the Sun first appear in the mid-latitudes, and over the course of a decade slowly creep toward the Sun's equator. Then, abruptly, the spots disappear, only to appear in the mid-latitudes again. While individual sunspots may only last a month or two, the collection of sunspots follows this regular cycle, and has done so for a very, very long time (probably for billions of years—though no one was around to see them).

Had it been observing our solar system, Kepler could have seen the evolution of the sunspot cycle. These spots are easy to identify in the Kepler data because of the brightness variations that they induce on the star. It is even possible to see where on the surface, at which latitudes, the sunspots appear. With each planet that transits the star, it is possible to determine whether it transits across the center or the limb of the star. If that planet also happens to transit across a star spot, it tells you where the star spots are found on the star. Even better is when the planetary orbit and the stellar orbit are misaligned by some angle. There you can see the small brightness changes that occur when the planet transits star spots in two different locations in the transit light curve. Those locations denote the north and south latitudes where the spots are concentrated. With PLATO data on Kepler stars, we will be able to see how those star spots, and other forms of stellar activity, evolve over timescales similar to their evolution on the Sun.

These missions, TESS and PLATO, are discovery missions where the goal is to find new transiting planetary systems and add them to the list of planets. The information they provide gives insights into the system architectures, formation histories, planet masses, sizes, and orbits, and the overall planet population. But they don't give many details about the planets themselves, their atmospheres, clouds, winds, or surface properties. Planet characterization takes different instruments, on different telescopes. A number of missions, campaigns, and instruments are being designed and built with this in mind. The CHEOPS mission (the CHaracterizing ExOPlanets Satellite) is a Swiss-led mission geared toward understanding the details of individual planets. And there are several missions and mission concepts that are being developed to measure the properties of exoplanet atmospheres, the James Webb Space Telescope featuring prominently in this arena.

We now know that the sizes of most of the Kepler planets place them between Earth and Neptune, and that we see two different types of planets: super-Earths that are rocky with essentially no atmosphere, and sub-Neptunes that retained an abundant atmosphere. However, the interior structure of these planets, especially sub-Neptunes, isn't entirely known. For example, it isn't yet clear what elements compose their atmospheres.

The atmospheres of Neptune and Uranus are made mostly of light hydrogen compounds—water, methane, and ammonia. But those molecules may not condense out of the hot protoplanetary disk where these sub-Neptune exoplanets are observed. Instead, it may be that their atmospheres are composed primarily of hydrogen and helium, which could be pulled directly from the gas disk, rather than accumulating from materials that condense and separate out of the gas and settle from there onto the planetary surfaces. If the atmospheres are mostly hydrogen and helium rather than methane or ammonia, then these intermediate-sized planets from Kepler (intermediate between gas giants and small, rocky planets) would be more similar to Saturn, which also has a rocky core with an extended atmosphere dominated by hydrogen and helium, than to Neptune with its atmosphere of more complex molecules.

This is the area where the James Webb Space Telescope will thrive. JWST is not a survey instrument, built to study large patches of the sky. It isn't built to find new planets the way that Kepler was. Instead, it can take those discoveries and unlock many new insights that Kepler and its successors can't provide [79]. Observations from these characterizing missions will showcase the variety of planet properties that exist in the galaxy—opening more ways to compare these distant worlds with our own. As amazing as each of these missions is, and as groundbreaking as their observations will be, they rely on the foundation of exoplanet discoveries made by Kepler and its successors.

Ever since the discovery of the first exoplanet, and especially with the large number of planets that we've seen in the subsequent few decades, how we view the Earth, its sibling planets, and our shared history will never be the same. The variety of newly found planets and planetary systems differs in many ways, both striking and subtle, from what occurred in the solar system. The thousands of planets discovered by the Kepler mission comprise the bulk of the known systems in our catalogs. Despite the differences between these Kepler systems and our solar system, they are not rare—nearly half of all stars host systems like these.

Rather than gas giants forming beyond the ice line, and staying in those outer regions where we thought they belonged, many gas giants find their way to small orbits under the intense heat of their host stars. These systems evolve primarily through eccentric migration, where their motion is excited into highly elliptical orbits by interactions with other planets or nearby stars. They then lose energy as the planet's interior heats up from the constant distortions they experience from the gravitational tidal forces of the host star. Eventually, they end up on tiny orbits ten times closer to their host star than Mercury is to the Sun. If this had happened in the solar system, we would not be here to tell the tale. All of the terrestrial planets, and some or all of the outer planets, would have been ejected, crashed into the Sun, or had their orbits disrupted by their troublesome sibling. Fortunately, that didn't happen.

A small fraction of these hot Jupiters migrate more slowly toward their stellar hosts. Had this happened to the solar system, the Earth (assuming it survived) would have been driven into an orbit of only one or a few days. The Earth's atmosphere would evaporate away, leaving nothing behind to breathe. There would be no atmosphere to carry energy from the daylit side of the planet to the nightside, no atmosphere to brighten the sky—leaving only the darkness of space to be seen when we look up, like the view we have from the Moon's surface. The Sun would loom so large it would fill a quarter of the sky—it's blazing white disk standing in stark contrast to the distant blackness.

On such a small orbit, the side of the Earth's surface facing the Sun would be a sea of molten rock, while the nightside would be a frozen wasteland. Or, had we been slightly smaller in mass, the Sun might not have stopped with just removing the atmosphere. It could have proceeded to start boiling away the rocky material of the mantle. Eventually, all that would remain in this scenario is an orbiting ball of metal. Fortunately, that didn't happen either. The Earth wasn't driven into a tiny, one-day orbit where the Sun would have boiled away what atmosphere we may have had, and melted one half of the planet while leaving the other half frozen. Yet systems like this are about as common as those hosting hot Jupiters—meaning that there are tens, or even hundreds, of millions of planets that suffered that fate.

What else didn't happen was that the solar system didn't form a tight collection of super-Earths or sub-Neptunes. Instead of producing the handful of small, terrestrial planets that are relatively isolated from their giant siblings, it apparently would have been easy to produce a set of planets twice the size and five times the mass of the Earth, all smooshed together on orbits interior to those of Mercury or Venus. The surface pressure from our atmosphere would far exceed the pressure in the depths of the ocean. The sky would be clouded over, and the energy that makes it down from the Sun and up from the Earth would be trapped at the surface by the atmosphere—a planet-sized pressure cooker. Hot Jupiters only happen in one out of a hundred systems. This possible fate was far more likely. Becoming a super-Earth with a thick atmosphere was only a coin toss away.

In these compact systems, strong dynamical interactions between planets are standard fare. If the Earth were to experience these effects, it could completely change a number of things that we take for granted. We could have had a nearby planetary neighbor, or two, whose gravitational influence would cause the length of the year to vary by a few weeks (or even a month) every decade or two. In this scenario, the 2020s might have thirteen months in each year and the 2030s only eleven. If nothing else, it would make it more challenging to schedule birthday parties. (One solution would be to have your birthday be more of a "birthday longitude," where you celebrate the moment when the Earth and Sun line up with the particular point in the sky where they were when you were born.) Regardless, it's probably a good thing that we don't have to address this issue. A stable orbital period helps with stabilizing other processes—like weather patterns.

Other effects would show up in these strongly interacting systems. In the solar system, the interplay between the Earth, Moon, and other planets generates cyclical changes to the Earth's orbital eccentricity, its obliquity, and its inclination. Adding stronger interactions between the Earth and the other planets could cause substantial changes to these parameters as well—not just to our orbital period. If the Earth's obliquity were to wander too much from the 23.5 degrees that it currently has, it could cause the Earth to completely freeze over, since the poles might not be exposed to enough sunlight to keep the ice at bay. As the ice advances, the Earth would be more reflective, and would cool even faster—accelerating the ice's advance. Or, if the obliquity were to flop around chaotically (which is also a possibility), then the climate wouldn't be stable enough to sustain a large population for a long period of time. Large swings in the amount of sunlight that a region receives could occur on timescales of a few hundred years.

Another insight we gain from the Kepler discoveries, which we cannot examine in the solar system, is the ability to look at long-term dynamical effects. The solar system is roughly five billion years old— meaning that the Earth has orbited about five billion times. Jupiter has only orbited the Sun one-tenth that number, and Uranus one-hundredth. Given the short orbital periods of many of the Kepler planets, there are many systems which are, dynamically speaking, significantly older than the solar system, since the age of a system is really measured by the number of orbits that the planets have completed.

A set of planets with orbital periods of a month or two have orbited their stellar host ten times more than the Earth has orbited the Sun. For those with orbits of a few days, that factor is over a hundred. The true dynamical ages of these systems are more like fifty or five hundred billion orbits, rather than the mere five billion count of the Earth. Such systems explore subtle dynamical effects that the solar system will not experience. The Sun will only live for ten billion years, and we are already halfway through that lifetime. We, on Earth, will not experience life in the solar system (as it currently is constituted) beyond ten billion orbits—let alone beyond a hundred billion. Many Kepler systems are already a hundred times that age in the dynamical sense.

Another situation we will not experience is the possibility of significant reservoirs of life on other, nearby planets. If the Earth had formed in a different system, perhaps a number of our planetary siblings would also live in the Sun's habitable zone, and have liquid water on their surfaces. Kepler uncovered several examples of potentially multihabitable systems where the conditions for life might be present on more than one planet. In the solar system, human habitation on multiple planets—like Mars, for example—will be limited to the point of virtual inconsequence. (A colony of a few thousand people on Mars or the Moon would only represent one-millionth of the population— perhaps an important millionth if some catastrophic collision were to take place on the Earth, but one-millionth nonetheless.)

However, in multihabitable systems, humanity or some spacefaring civilization could potentially flourish in large numbers on multiple planets. In the case of the TRAPPIST system, the number of potentially habitable planets is at least three, and could be four. It isn't hard to imagine that life could spread from one planet to another, either naturally through material ejected from an asteroid collision, or artificially by an intelligent species making the journey on a craft of some sort. And, in this case, there may not be a need for small, hermetically sealed, and artificial habitats to keep them alive at their destination.

Another thing to consider is how different life on Earth would be if we had a different Sun. If the Earth had formed around a random star, the chances are that the star would have been a small M-dwarf star. M-dwarfs are by far the most common type of star—more than twenty times as abundant as stars like the Sun. Even if the definition of Sun-like stars is expanded from G-type stars to include the neighboring F- (slightly larger) and K-type (slightly smaller) stars, M-dwarfs are still five times more abundant. From the Kepler data, we see that small, Earth-sized planets are more common around these types of stars than they are around stars like the Sun, so not only are M-dwarf stars more abundant, but there are more planets for each one.

While random chance would have favored the Earth forming around one of these stars, orbiting an M-dwarf star would be an entirely different experience from what we know. The stars themselves are less than half the size and mass of the Sun, and their surface temperatures are about half as well. The light we would receive would only be one percent of what we receive from the Sun—and it would be brightest in the infrared rather than the visible part of the spectrum. That could affect a whole host of evolutionary processes, such as how plants convert the starlight into energy, or how our eyes work and what colors we would perceive.

Life around an M-dwarf would also be foreign because of the variety of magnetically driven effects. Like the Sun, M-dwarfs also have spots, lots of them, and compared to the size of the star, they're big. These star spots cause significant changes in the brightness of the star when viewed from the orbiting planets. During a period of high activity, when the face of the star is mottled by several star spots, the total energy received by the planets can drop by twenty-five percent, or more. The star-spot cycle could persist for years (like the Sun's eleven-year cycle) with the brightness of the star fluctuating by a significant fraction whenever the spots come into or out of view.

The weather patterns resulting from these large variations in energy received from the star would be strange, with sudden changes in wind and rain cycles over a decade-long timescale as the star's brightness varies from its spots. On the Earth, there are some regions where the weather is dominated by the multiyear La Niña and El Niño cycles, rather than the yearly seasonal cycles. Australia is one example. The durations of these unpredictable cycles can range from a few to several years and are driven by ocean surface temperatures, accompanied by changes in the locations of atmospheric high- and low-pressure centers over the Pacific. For a planet orbiting an M-dwarf, weather variations would be driven by the fact that a quarter of the star's light is attenuated as the spots grow in number and size.

These weather differences don't mention the direct effects of stellar activity on the planet. As the spots on the Sun increase in number, they are accompanied by large solar storms, flares, and bursts of energetic charged particles in coronal mass ejections. This space weather affects not only the Earth's atmosphere (causing flares in the northern lights, and compressing the atmosphere's outermost layers), but it can also affect electronic devices like satellites and communications systems. We saw the effects of several coronal mass ejections on the electronics of Kepler. The same kinds of events would happen with an M-dwarf, affecting electronic devices in orbit around the star or planet, and on the planetary surface. But now, in the habitable zone of one of these smaller stars, you would be ten times closer—making the effects a hundred times more intense.

These challenges for a civilization living on an M-dwarf planet, as it relates to the low-intensity and highly variable light coming from the star, might make it an undesirable place to live. But planets orbiting M-stars do come with at least one significant advantage: M-stars have exceptionally long lifetimes. We are currently circling a star that is halfway through its ten-billion-year lifetime. But, if we were to orbit a small M-dwarf star, born on the same day as the Sun, we would currently have more than ninety-nine percent of that star's life ahead of us. The smallest M-dwarf stars can live for over a trillion years—ten times longer than the current age of the universe.

Their longevity comes from the fact that they burn through their fuel so slowly. While they may be ten percent of the mass of the Sun, they are thousands of times dimmer, slowly consuming their hydrogen reserves over eons. The same cannot be said of stars more massive than the Sun. The more massive the star in a system, the hotter the core becomes and the faster the star converts its hydrogen into helium and energy. Stars that are a few times the mass of the Sun might only live for a hundred million years. That might sound like a long time, but it is shorter than the amount of time the dinosaurs roamed the Earth. It's also shorter than the amount of time that it takes to form some planets. Indeed, some high-mass stars have lifetimes that are so short that planets simply never have time to coalesce out of the nebula. Or, if they do, they don't have much time before the star starts preparing for its demise.

One thing we see with our observations is that if there is time for planets to form, then they generally will. Planets indeed form around massive stars with relatively short lifetimes. These A-type stars, are perhaps two solar masses to begin with, and live roughly one billion years—or one-tenth the lifetime of the Sun. For Kepler's purposes, a Sun-like star has a mass between roughly a half and one and a half times the mass of the Sun. These A-stars are more massive than this amount, and larger in diameter as a consequence, so a transiting planet blocks a smaller portion of the starlight. Nevertheless, Kepler data show planets orbiting these kinds of stars.

Though A-stars were not the primary focus of the mission, their inclusion provides some important insights, because A-type stars present several challenges for almost all methods of planet detection. They are much hotter than Sun-like stars—nearly 10,000 kelvin, compared to our 5800 kelvin Sun, and these high temperatures reduce the number of spectral lines we observe in the outer layers of the stellar atmosphere. They also spin rapidly, because their lack of a magnetic field limits the braking that occurs for smaller stars. These effects make Doppler detection of other planets difficult, so the Kepler data are particularly valuable. Those data show that these stars have planets, and because these stars are such different animals from the Sun, Kepler provided unique insights into the nature of planets in their hot, short-lived environments.

Given the number of surprises that exoplanet discoveries have produced, from compact systems of sub-Neptune planets, to planets with tails of evaporated rocks, to planets orbiting both stars in a binary pair, if there is anything that the Kepler mission showed us, it was that there remains a lot for us to learn about the Earth's distant cousins. After more than three decades, from its conception in the 1980s to its demise in the 2010s, this brainchild of William Borucki should claim a well-deserved spot in the pantheon of scientific endeavors.

As with any human enterprise, the mission and its people had their share of setbacks, missed promises, compromises, scandal, infighting, and loss. We lost one of the most important figures in the fall of 2012 with the death of David Koch, the Deputy Principal Investigator and the person who developed the stretched-wire test that demonstrated Kepler's potential and which broke the final barrier to getting Kepler approved by NASA [80]. More than its challenges, however, Kepler had its share of triumphs, breakthroughs, and brilliant results.

The mission stands as a testament to the noble fragility of the human condition. For centuries, mankind has looked at the heavens in wonder, considering the possibility of other distant planets with potential distant denizens. Our ancestors were pondering these questions while scratching their existence from the thin, living surface of a small rock, orbiting a nondescript star. While Kepler did not answer the ultimate question about our collective solitude in the universe, it clearly showed the wondrous variety of creation that is out there.

As we explore these strange new worlds with new missions and new instrumentation, it's worth remembering the pioneering efforts that laid the foundation of the discipline. Kepler was unprecedented in the sheer volume of discoveries that it made. Many of its discoveries ran counter to our prevailing understanding of the origins of planets and planetary systems. Incorporating these results into a more expansive, or more general, theory of our origins is a work in progress—but there is real progress.

As Johannes Kepler said, centuries ago, "The treasures hidden in the heavens are so rich that the human mind shall never be lacking in fresh nourishment." In the coming years, and coming decades, the results of his namesake mission will undoubtedly be at the center of advances in the field of exoplanets. They will serve as a source for the nucleation of new ideas and as a launchpad for new investigations. In a broader sense, Kepler will also be at the center of our advances in how we understand our own solar system, and our own place in the universe.

BIBLIOGRAPHY

[1] M. Mayor and D. Queloz. A Jupiter-Mass Companion to a Solar-Type Star. *Nature*, 378(6555):355–359, November 1995.

[2] O. Struve. Proposal for a Project of High-Precision Stellar Radial Velocity Work. *Observatory*, 72:199–200, October 1952.

[3] W. J. Borucki and A. L. Summers. The Photometric Method of Detecting Other Planetary Systems. *Icarus*, 58(1):121–134, April 1984.

[4] W. J. Borucki, E. W. Dunham, D. G. Koch, W. D. Cochran, J. D. Rose, D. K. Cullers, A. Granados, and J. M. Jenkins. FRESIP: A Mission to Determine the Character and Frequency of Extra-Solar Planets around Solar-Like Stars. *APSS*, 241(1):111–134, March 1996.

[5] W. J. Borucki. KEPLER Mission: Development and Overview. *Reports on Progress in Physics*, 79(3):036901, March 2016.

[6] A. Wolszczan and D. A. Frail. A Planetary System around the Millisecond Pulsar PSR1257 + 12. *Nature*, 355(6356):145–147, January 1992.

[7] E. Agol, J. Steffen, R. Sari, and W. Clarkson. On Detecting Terrestrial Planets with Timing of Giant Planet Transits. *MNRAS*, 359(2):567–579, May 2005.

[8] M. J. Holman and N. W. Murray. The Use of Transit Timing to Detect Terrestrial-Mass Extrasolar Planets. *Science*, 307(5713):1288–1291, February 2005.

[9] J. Miralda-Escudé. Orbital Perturbations of Transiting Planets: A Possible Method to Measure Stellar Quadrupoles and to Detect Earth-Mass Planets. *ApJ*, 564(2):1019–1023, January 2002.

[10] J. Steffen. *Detecting New Planets in Transiting Systems*. PhD thesis, University of Washington, Seattle, January 2006.

[11] K. Batygin and M. E. Brown. Evidence for a Distant Giant Planet in the Solar System. *AJ*, 151(2):22, February 2016.

[12] M. E. Brown, C. A. Trujillo, and D. L. Rabinowitz. Discovery of a Planetary-Sized Object in the Scattered Kuiper Belt. *ApJL*, 635(1):L97–L100, December 2005.

[13] Flavien Kiefer. Determining the mass of the planetary candidate HD 114762 b using Gaia. *AAP*, 632:L9, December 2019.

[14] D. Charbonneau, T. M. Brown, D. W. Latham, and M. Mayor. Detection of Planetary Transits across a Sun-Like Star. *ApJL*, 529(1):L45–L48, January 2000.

[15] G. W. Henry, G. W. Marcy, R. Paul Butler, and S. S. Vogt. A Transiting "51 Peg-Like" Planet. *ApJL*, 529(1):L41–L44, January 2000.

[16] S. B. Howell. *The NASA Kepler Mission*, IOP Publishing, Bristol, UK, 2020.

[17] T. M. Brown, D. W. Latham, M. E. Everett, and G. A. Esquerdo. Kepler Input Catalog: Photometric Calibration and Stellar Classification. *AJ*, 142(4):112, October 2011.

[18] NASA. Kepler Mission Participating Scientists. https://nspires.nasaprs.com/external/solicitations/summary!init.do?solId=0B53C12358C969F7D74B9E7A1A5DD8E1, 2007.

[19] NASA. NASA Releases Orbiting Carbon Observatory Accident Summary. https://www.jpl.nasa.gov/news/nasa-releases-orbiting-carbon-observatory-accident-summary, 2009.

[20] NASA. Kepler Data Release Notes. https://archive.stsci.edu/missions-and-data/kepler/documents/data-release-notes, 2018.

[21] W. J. Borucki, D. G. Koch, T. M. Brown, G. Basri, N. M. Batalha, D. A. Caldwell, W. D. Cochran, E. W. Dunham, T. N. Gautier III, J. C. Geary, R. L. Gilliland, S. B. Howell, J. M. Jenkins, D. W. Latham, J. J. Lissauer, G. W. Marcy, D. Monet, J. F. Rowe, and D. Sasselov. Kepler-4b: A Hot Neptune-Like Planet of a G0 Star near Main-Sequence Turnoff. *ApJL*, 713(2):L126–L130, April 2010.

[22] D. G. Koch, W. J. Borucki, J. F. Rowe, N. M. Batalha, T. M. Brown, D. A. Caldwell, J. Caldwell, W. D. Cochran, E. DeVore, E. W. Dunham, A. K. Dupree, T. N. Gautier III, J. C. Geary, R. L. Gilliland, S. B. Howell, J. M. Jenkins, D. W. Latham, J. J. Lissauer, G. W. Marcy, D. Morrison, and J. Tarter. Discovery of the Transiting Planet Kepler-5b. *ApJL*, 713(2):L131–L135, April 2010.

[23] E. W. Dunham, W. J. Borucki, D. G. Koch, N. M. Batalha, L. A. Buchhave, T. M. Brown, D. A. Caldwell, W. D. Cochran, M. Endl, D. Fischer, G. Fűrész, T. N. Gautier III, J. C. Geary, R. L. Gilliland, A. Gould, S. B. Howell, J. M. Jenkins, H. Kjeldsen, D. W. Latham, J. J. Lissauer, G. W. Marcy, S. Meibom, D. G. Monet, J. F. Rowe, and D. D. Sasselov. Kepler-6b: A Transiting Hot Jupiter Orbiting a Metal-Rich Star. *ApJL*, 713(2):L136–L139, April 2010.

[24] J. M. Jenkins, W. J. Borucki, D. G. Koch, G. W. Marcy, W. D. Cochran, W. F. Welsh, G. Basri, N. M. Batalha, L. A. Buchhave, T. M. Brown, D. A. Caldwell, E. W. Dunham, M. Endl, D. A. Fischer, T. N. Gautier III, J. C. Geary, R. L. Gilliland, S. B. Howell, H. Isaacson, J. A. Johnson, D. W. Latham, J. J. Lissauer, D. G. Monet, J. F. Rowe, D. D. Sasselov, A. W. Howard, P. MacQueen, J. A. Orosz, H. Chandrasekaran, J. D. Twicken, S. T. Bryson, E. V. Quintana, B. D. Clarke, J. Li, C. Allen, P. Tenenbaum, H. Wu, S. Meibom, T. C. Klaus, C. K. Middour, M. T. Cote, S. McCauliff, F. R. Girouard, J. P. Gunter, B. Wohler, J. R. Hall, K. Ibrahim, A. Kamal Uddin, M. S. Wu, P. A. Bhavsar, J. Van Cleve, D. L. Pletcher, J. L. Dotson, and M. R. Haas. Discovery and Rossiter–McLaughlin Effect of Exoplanet Kepler-8b. *ApJ*, 724(2):1108–1119, December 2010.

[25] D. W. Latham, W. J. Borucki, D. G. Koch, T. M. Brown, L. A. Buchhave, G. Basri, N. M. Batalha, D. A. Caldwell, W. D. Cochran, E. W. Dunham, G. Fűrész, T. N. Gautier III, J. C. Geary, R. L. Gilliland, S. B. Howell, J. M. Jenkins, J. J. Lissauer, G. W. Marcy, D. G. Monet, J. F. Rowe, and D. D. Sasselov. Kepler-7b: A Transiting Planet with Unusually Low Density. *ApJL*, 713(2):L140–L144, April 2010.

[26] W. J. Borucki, D. Koch, J. Jenkins, D. Sasselov, R. Gilliland, N. Batalha, D. W. Latham, D. Caldwell, G. Basri, T. Brown, J. Christensen-Dalsgaard, W. D. Cochran, E. DeVore, E. Dunham, A. K. Dupree, T. Gautier, J. Geary, A. Gould, S. Howell, H. Kjeldsen, J. Lissauer, G. Marcy, S. Meibom, D. Morrison, and J. Tarter. Kepler's Optical Phase Curve of the Exoplanet HAT-P-7b. *Science*, 325(5941):709, August 2009.

[27] W. F. Welsh, J. A. Orosz, S. Seager, J. J. Fortney, J. Jenkins, J. F. Rowe, D. Koch, and W. J. Borucki. The Discovery of Ellipsoidal Variations in the Kepler Light Curve of HAT-P-7. *ApJL*, 713(2):L145–L149, April 2010.

[28] E. Hand. Telescope Team May Be Allowed to Sit on Exoplanet Data. *Nature*, 2010.

[29] A. Léger, D. Rouan, J. Schneider, P. Barge, M. Fridlund, B. Samuel, M. Ollivier, E. Guenther, M. Deleuil, H. J. Deeg, M. Auvergne, R. Alonso, S. Aigrain, A. Alapini, J. M. Almenara, A. Baglin, M. Barbieri, H. Bruntt, P. Bordé, F. Bouchy, J. Cabrera, C. Catala, L. Carone, S. Carpano, Sz. Csizmadia, R. Dvorak, A. Erikson, S. Ferraz-Mello, B. Foing, F. Fressin, D. Gandolfi, M. Gillon, Ph. Gondoin, O. Grasset, T. Guillot, A. Hatzes, G. Hébrard, L. Jorda, H. Lammer, A. Llebaria, B. Loeillet, M. Mayor, T. Mazeh, C. Moutou, M. Pätzold, F. Pont, D. Queloz, H. Rauer, S. Renner, R. Samadi, A. Shporer, Ch. Sotin, B. Tingley, G. Wuchterl, M. Adda, P. Agogu, T. Appourchaux, H. Ballans, P. Baron, T. Beaufort, R. Bellenger, R. Berlin, P. Bernardi, D. Blouin, F. Baudin, P. Bodin, L. Boisnard, L. Boit, F. Bonneau, S. Borzeix, R. Briet, J. T. Buey, B. Butler, D. Cailleau, R. Cautain, P. Y. Chabaud, S. Chaintreuil, F. Chiavassa, V. Costes, V. Cuna Parrho, F. de Oliveira Fialho, M. Decaudin, J. M. Defise, S. Djalal, G. Epstein, G. E. Exil, C. Fauré, T. Fenouillet, A. Gaboriaud, A. Gallic, P. Gamet, P. Gavalda, E. Grolleau, R. Gruneisen, L. Gueguen, V. Guis, V. Guivarc'h, P. Guterman, D. Hallouard, J. Hasiba, F. Heuripeau, G. Huntzinger, H. Hustaix, C. Imad, C. Imbert, B. Johlander, M. Jouret, P. Journoud, F. Karioty, L. Kerjean, V. Lafaille, L. Lafond, T. Lam-Trong, P. Landiech, V. Lapeyrere, T. Larqué, P. Laudet, N. Lautier, H. Lecann, L. Lefevre, B. Leruyet, P. Levacher, A. Magnan, E. Mazy, F. Mertens, J. M. Mesnager, J. C. Meunier, J. P. Michel, W. Monjoin, D. Naudet, K. Nguyen-Kim, J. L. Orcesi, H. Ottacher, R. Perez, G. Peter, P. Plasson, J. Y. Plesseria, B. Pontet, A. Pradines, C. Quentin, J. L. Reynaud, G. Rolland, F. Rollenhagen, R. Romagnan, N. Russ, R. Schmidt, N. Schwartz, I. Sebbag, G. Sedes, H. Smit, M. B. Steller, W. Sunter, C. Surace, M. Tello, D. Tiphène, P. Toulouse, B. Ulmer, O. Vandermarcq, E. Vergnault, A. Vuillemin, and P. Zanatta. Transiting Exoplanets from the CoRoT Space Mission. VIII. CoRoT-7b: The First Super-Earth with Measured Radius. *AAP*, 506(1):287–302, October 2009.

[30] J. H. Steffen, D. Ragozzine, D. C. Fabrycky, J. A. Carter, E. B. Ford, M. J. Holman, J. F. Rowe, W. F. Welsh, W. J. Borucki, A. P. Boss, D. R. Ciardi, and S. N. Quinn. Kepler Constraints on Planets near Hot Jupiters. *Proceedings of the National Academy of Science*, 109(21):7982–7987, May 2012.

[31] Y.-C. Hong, S. N. Raymond, P. D. Nicholson, and J. I. Lunine. Innocent Bystanders: Orbital Dynamics of Exomoons during Planet–Planet Scattering. *ApJ*, 852(2):85, January 2018.

[32] I. Rabago and J. H. Steffen. Survivability of Moon Systems around Ejected Gas Giants. *MNRAS*, 489(2):2323–2329, October 2019.

[33] B. E. McArthur, M. Endl, W. D. Cochran, G. F. Benedict, D. A. Fischer, G. W. Marcy, R. P. Butler, D. Naef, M. Mayor, D. Queloz, S. Udry, and T. E. Harrison. Detection of a Neptune-Mass Planet in the ρ^1 Cancri System Using the Hobby–Eberly Telescope. *ApJL*, 614(1):L81–L84, October 2004.

[34] D. A. Fischer, G. W. Marcy, R. P. Butler, S. S. Vogt, G. Laughlin, G. W. Henry, D. Abouav, K.M.G. Peek, J. T. Wright, J. A. Johnson, C. McCarthy, and H. Isaacson. Five Planets Orbiting 55 Cancri. *ApJ*, 675(1):790–801, March 2008.

[35] Joshua N. Winn, Jaymie M. Matthews, Rebekah I. Dawson, Daniel Fabrycky, Matthew J. Holman, Thomas Kallinger, Rainer Kuschnig, Dimitar Sasselov, Diana Dragomir, David B. Guenther, Anthony F. J. Moffat, Jason F. Rowe, Slavek Rucinski, and Werner W. Weiss. A Super-Earth Transiting a Naked-eye Star. *ApJL*, 737(1):L18, August 2011.

[36] R. I. Dawson and D. C. Fabrycky. Radial Velocity Planets De-aliased: A New, Short Period for Super-Earth 55 Cnc e. *ApJ*, 722(1):937–953, October 2010.

[37] S. Rappaport, A. Levine, E. Chiang, I. El Mellah, J. Jenkins, B. Kalomeni, E. S. Kite, M. Kotson, L. Nelson, L. Rousseau-Nepton, and K. Tran. Possible Disintegrating Short-Period Super-Mercury Orbiting KIC 12557548. *ApJ*, 752(1):1, June 2012.

[38] J. N. Winn, R. Sanchis-Ojeda, and S. Rappaport. Kepler-78 and the Ultra-Short-Period Planets. *New Astron. Rev.*, 83:37–48, November 2018.

[39] F. Valsecchi, F. A. Rasio, and J. H. Steffen. From Hot Jupiters to Super-Earths via Roche Lobe Overflow. *ApJL*, 793(1):L3, September 2014.

[40] H. A. Knutson, D. Charbonneau, L. E. Allen, J. J. Fortney, E. Agol, N. B. Cowan, A. P. Showman, C. S. Cooper, and S. T. Megeath. A Map of the Day–Night Contrast of the Extrasolar Planet HD 189733b. *Nature*, 447(7141):183–186, May 2007.

[41] D. J. Armstrong, T. A. Lopez, V. Adibekyan, R. A. Booth, E. M. Bryant, K. A. Collins, M. Deleuil, A. Emsenhuber, C. X. Huang, G. W. King, J. Lillo-Box, J. J. Lissauer, E. Matthews, O. Mousis, L. D. Nielsen, H. Osborn, J. Otegi, N. C. Santos, S. G. Sousa, K. G. Stassun, D. Veras, C. Ziegler, J. S. Acton, J. M. Almenara, D. R. Anderson, D. Barrado, S.C.C. Barros, D. Bayliss, C. Belardi, F. Bouchy, C. Briceño, M. Brogi, D.J.A. Brown, M. R. Burleigh, S. L. Casewell, A. Chaushev, D. R. Ciardi, K. I. Collins, K. D. Colón, B. F. Cooke, I.J.M. Crossfield, R. F. Díaz, E. Delgado Mena, O.D.S. Demangeon, C. Dorn, X. Dumusque, P. Eigmüller, M. Fausnaugh, P. Figueira, T. Gan, S. Gandhi, S. Gill, E. J. Gonzales, M. R. Goad, M. N. Günther, R. Helled, S. Hojjatpanah, S. B. Howell, J. Jackman, J. S. Jenkins, J. M. Jenkins, E.L.N. Jensen, G. M. Kennedy, D. W. Latham, N. Law, M. Lendl, M. Lozovsky, A. W. Mann, M. Moyano, J. McCormac, F. Meru, C. Mordasini, A. Osborn, D. Pollacco, D. Queloz, L. Raynard, G. R. Ricker, P. Rowden, A. Santerne, J. E. Schlieder, S. Seager, L. Sha, T.-G. Tan, R. H. Tilbrook, E. Ting, S. Udry, R. Vanderspek, C. A. Watson, R. G. West, P. A. Wilson, J. N. Winn, P. Wheatley, J. N. Villasenor, J. I. Vines, and Z. Zhan. A Remnant Planetary Core in the Hot-Neptune Desert. *Nature*, 583(7814):39–42, July 2020.

[42] J. H. Steffen, N. M. Batalha, W. J. Borucki, L. A. Buchhave, D. A. Caldwell, W. D. Cochran, M. Endl, D. C. Fabrycky, F. Fressin, E. B. Ford, J. J. Fortney, M. J. Haas, M. J. Holman, S. B. Howell, H. Isaacson, J. M. Jenkins, D. Koch, D. W. Latham, J. J. Lissauer, A. V. Moorhead, R. C. Morehead, G. Marcy, P. J. MacQueen, S. N. Quinn, D. Ragozzine, J. F. Rowe, D. D. Sasselov, S. Seager, G. Torres, and W. F. Welsh. Five Kepler Target Stars That Show Multiple Transiting Exoplanet Candidates. *ApJ*, 725(1):1226–1241, December 2010.

[43] M. J. Holman, D. C. Fabrycky, D. Ragozzine, E. B. Ford, J. H. Steffen, W. F. Welsh, J. J. Lissauer, D. W. Latham, G. W. Marcy, L. M. Walkowicz, N. M. Batalha, J. M. Jenkins, J. F. Rowe, W. D. Cochran, F. Fressin, G. Torres, L. A. Buchhave, D. D. Sasselov, W. J. Borucki, D. G. Koch, G. Basri, T. M. Brown, D. A. Caldwell, D. Charbonneau, E. W. Dunham, T. N. Gautier, J. C. Geary, R. L. Gilliland, M. R. Haas, S. B. Howell, D. R. Ciardi, M. Endl, D. Fischer, G. Fűrész, J. D. Hartman, H. Isaacson, J. A. Johnson, P. J. MacQueen, A. V. Moorhead, R. C. Morehead, and J. A. Orosz. Kepler-9: A System of Multiple Planets Transiting a Sun-Like Star, Confirmed by Timing Variations. *Science*, 330(6000):51, October 2010.

[44] J. J. Lissauer, D. C. Fabrycky, E. B. Ford, W. J. Borucki, F. Fressin, G. W. Marcy, J. A. Orosz, J. F. Rowe, G. Torres, W. F. Welsh, N. M. Batalha, S. T. Bryson, L. A. Buchhave, D. A. Caldwell, J. A. Carter, D. Charbonneau, J. L. Christiansen, W. D. Cochran, J.-M. Desert, E. W. Dunham, M. N. Fanelli, J. J. Fortney, T. N. Gautier III, J. C. Geary, R. L. Gilliland, M. R. Haas, J. R. Hall, M. J. Holman, D. G. Koch, D. W. Latham, E. Lopez, S. McCauliff, N. Miller, R. C. Morehead, E. V. Quintana, D. Ragozzine, D. Sasselov, D. R. Short, and J. H. Steffen. A Closely Packed System of Low-Mass, Low-Density Planets Transiting Kepler-11. *Nature*, 470(7332):53–58, February 2011.

[45] J. A. Carter, E. Agol, W. J. Chaplin, S. Basu, T. R. Bedding, L. A. Buchhave, J. Christensen-
      Dalsgaard, K. M. Deck, Y. Elsworth, D. C. Fabrycky, E. B. Ford, J. J. Fortney, S. J. Hale,
      R. Handberg, S. Hekker, M. J. Holman, D. Huber, C. Karoff, S. D. Kawaler, H. Kjeldsen,
      J. J. Lissauer, E. D. Lopez, M. N. Lund, M. Lundkvist, T. S. Metcalfe, A. Miglio, L. A.
      Rogers, D. Stello, W. J. Borucki, S. Bryson, J. L. Christiansen, W. D. Cochran, J. C. Geary,
      R. L. Gilliland, M. R. Haas, J. Hall, A. W. Howard, J. M. Jenkins, T. Klaus, D. G. Koch,
      D. W. Latham, P. J. MacQueen, D. Sasselov, J. H. Steffen, J. D. Twicken, and J. N. Winn.
      Kepler-36: A Pair of Planets with Neighboring Orbits and Dissimilar Densities. *Science*,
      337(6094):556, August 2012.
[46] K. M. Deck, M. J. Holman, E. Agol, J A. Carter, J. J. Lissauer, D. Ragozzine, and J. N.
      Winn. Rapid Dynamical Chaos in an Exoplanetary System. *ApJL*, 755(1):L21, August
      2012.
[47] J. Laskar and M. Gastineau. Existence of Collisional Trajectories of Mercury, Mars and
      Venus with the Earth. *Nature*, 459(7248):817–819, June 2009.
[48] S. M. Mills, D. C. Fabrycky, C. Migaszewski, E. B. Ford, E. Petigura, and H. Isaacson. A
      Resonant Chain of Four Transiting, Sub-Neptune Planets. *Nature*, 533(7604):509–512,
      May 2016.
[49] M. G. MacDonald, D. Ragozzine, D. C. Fabrycky, E. B. Ford, M. J. Holman, H. T. Isaac-
      son, J. J. Lissauer, E. D. Lopez, T. Mazeh, L. Rogers, J. F. Rowe, J. H. Steffen, and G.
      Torres. A Dynamical Analysis of the Kepler-80 System of Five Transiting Planets. *AJ*,
      152(4):105, October 2016.
[50] M. Gillon, A.H.M.J. Triaud, B.-O. Demory, E. Jehin, E. Agol, K. M. Deck, S. M. Lederer,
      J. de Wit, A. Burdanov, J. G. Ingalls, E. Bolmont, J. Leconte, S. N. Raymond, F. Selsis,
      M. Turbet, K. Barkaoui, A. Burgasser, M. R. Burleigh, S. J. Carey, A. Chaushev, C. M.
      Copperwheat, L. Delrez, C. S. Fernandes, D. L. Holdsworth, E. J. Kotze, V. Van Grootel,
      Y. Almleaky, Z. Benkhaldoun, P. Magain, and D. Queloz. Seven Temperate Terrestrial
      Planets around the Nearby Ultracool Dwarf Star TRAPPIST-1. *Nature*, 542(7642):
      456–460, February 2017.
[51] R. Luger, M. Sestovic, E. Kruse, S. L. Grimm, B.-O. Demory, E. Agol, E. Bolmont, D. Fab-
      rycky, C. S. Fernandes, V. Van Grootel, A. Burgasser, M. Gillon, J. G. Ingalls, E. Jehin, S. N.
      Raymond, F. Selsis, A.H.M.J. Triaud, T. Barclay, G. Barentsen, S. B. Howell, L. Delrez, J.
      de Wit, D. Foreman-Mackey, D. L. Holdsworth, J. Leconte, S. Lederer, M. Turbet, Y. Alm-
      leaky, Z. Benkhaldoun, P. Magain, B. M. Morris, K. Heng, and D. Queloz. A Seven-Planet
      Resonant Chain in TRAPPIST-1. *Nature Astronomy*, 1:0129, June 2017.
[52] E. Kruse and E. Agol. KOI-3278: A Self-Lensing Binary Star System. *Science*,
      344(6181):275–277, April 2014.
[53] W. F. Welsh, J. A. Orosz, C. Aerts, T. M. Brown, E. Brugamyer, W. D. Cochran, R. L.
      Gilliland, J. A. Guzik, D. W. Kurtz, D. W. Latham, G. W. Marcy, S. N. Quinn, W. Zima,
      C. Allen, N. M. Batalha, S. Bryson, L. A. Buchhave, D. A. Caldwell, T. N. Gautier III,
      S. B. Howell, K. Kinemuchi, K. A. Ibrahim, H. Isaacson, J. M. Jenkins, A. Prsa, M.
      Still, R. Street, B. Wohler, D. G. Koch, and W. J. Borucki. KOI-54: The Kepler Discov-
      ery of Tidally Excited Pulsations and Brightenings in a Highly Eccentric Binary. *ApJS*,
      197(1):4, November 2011.
[54] R. L. Gilliland, T. M. Brown, J. Christensen-Dalsgaard, H. Kjeldsen, C. Aerts, T. Appour-
      chaux, S. Basu, T. R. Bedding, W. J. Chaplin, M. S. Cunha, P. De Cat, J. De Ridder,
      J. A. Guzik, G. Handler, S. Kawaler, L. Kiss, K. Kolenberg, D. W. Kurtz, T. S. Metcalfe,
      M.J.P.F.G. Monteiro, R. Szabó, T. Arentoft, L. Balona, J. Debosscher, Y. P. Elsworth,
      P.-O. Quirion, D. Stello, J. C. Suárez, W. J. Borucki, J. M. Jenkins, D. Koch, Y. Kondo, D. W.
      Latham, J. F. Rowe, and J. H. Steffen. Kepler Asteroseismology Program: Introduction
      and First Results. *PASP*, 122(888):131, February 2010.

[55] S. Meibom, S. A. Barnes, D. W. Latham, N. Batalha, W. J. Borucki, D. G. Koch, G. Basri, L. M. Walkowicz, K. A. Janes, J. Jenkins, J. Van Cleve, M. R. Haas, S. T. Bryson, A. K. Dupree, G. Furesz, A. H. Szentgyorgyi, L. A. Buchhave, B. D. Clarke, J. D. Twicken, and E. V. Quintana. The Kepler Cluster Study: Stellar Rotation in NGC 6811. *ApJL*, 733(1):L9, May 2011.

[56] A. McQuillan, T. Mazeh, and S. Aigrain. Rotation Periods of 34,030 Kepler Main-Sequence Stars: The Full Autocorrelation Sample. *ApJS*, 211(2):24, April 2014.

[57] J. J. Lissauer, G. W. Marcy, S. T. Bryson, J. F. Rowe, D. Jontof-Hutter, E. Agol, W. J. Borucki, J. A. Carter, E. B. Ford, R. L. Gilliland, R. Kolbl, K. M. Star, J. H. Steffen, and G. Torres. Validation of Kepler's Multiple Planet Candidates. II. Refined Statistical Framework and Descriptions of Systems of Special Interest. *ApJ*, 784(1):44, March 2014.

[58] L. R. Doyle, J. A. Carter, D. C. Fabrycky, R. W. Slawson, S. B. Howell, J. N. Winn, J. A. Orosz, A. Přsa, W. F. Welsh, S. N. Quinn, D. Latham, G. Torres, L. A. Buchhave, G. W. Marcy, J. J. Fortney, A. Shporer, E. B. Ford, J. J. Lissauer, D. Ragozzine, M. Rucker, N. Batalha, J. M. Jenkins, W. J. Borucki, D. Koch, C. K. Middour, J. R. Hall, S. McCauliff, M. N. Fanelli, E. V. Quintana, M. J. Holman, D. A. Caldwell, M. Still, R. P. Stefanik, W. R. Brown, G. A. Esquerdo, S. Tang, G. Furesz, J. C. Geary, P. Berlind, M. L. Calkins, D. R. Short, J. H. Steffen, D. Sasselov, E. W. Dunham, W. D. Cochran, A. Boss, M. R. Haas, D. Buzasi, and D. Fischer. Kepler-16: A Transiting Circumbinary Planet. *Science*, 333(6049):1602, September 2011.

[59] J. L. Smallwood, R. Nealon, C. Chen, R. G. Martin, J. Bi, R. Dong, and C. Pinte. GW Ori: Circumtriple Rings and Planets. *MNRAS*, 508(1):392–407, November 2021.

[60] K. Tsiganis, R. Gomes, A. Morbidelli, and H. F. Levison. Origin of the Orbital Architecture of the Giant Planets of the Solar System. *Nature*, 435(7041):459–461, May 2005.

[61] S. E. Thompson, J. L. Coughlin, K. Hoffman, F. Mullally, J. L. Christiansen, C. J. Burke, S. Bryson, N. Batalha, M. R. Haas, J. Catanzarite, J. F. Rowe, G. Barentsen, D. A. Caldwell, B. D. Clarke, J. M. Jenkins, J. Li, D. W. Latham, J. J. Lissauer, S. Mathur, R. L. Morris, S. E. Seader, J. C. Smith, T. C. Klaus, J. D. Twicken, J. E. Van Cleve, B. Wohler, R. Akeson, D. R. Ciardi, W. D. Cochran, C. E. Henze, S. B. Howell, D. Huber, A. Prša, S. V. Ramírez, T. D. Morton, T. Barclay, J. R. Campbell, W. J. Chaplin, D. Charbonneau, J. Christensen-Dalsgaard, J. L. Dotson, L. Doyle, E. W. Dunham, A. K. Dupree, E. B. Ford, J. C. Geary, F. R. Girouard, H. Isaacson, H. Kjeldsen, E. V. Quintana, D. Ragozzine, M. Shabram, A. Shporer, V. Silva Aguirre, J. H. Steffen, M. Still, P. Tenenbaum, W. F. Welsh, A. Wolfgang, K. A. Zamudio, D. G. Koch, and W. J. Borucki. Planetary Candidates Observed by Kepler. VIII. A Fully Automated Catalog with Measured Completeness and Reliability Based on Data Release 25. *ApJS*, 235(2):38, April 2018.

[62] J. L. Christiansen, B. D. Clarke, C. J. Burke, J. M. Jenkins, T. S. Barclay, E. B. Ford, M. R. Haas, A. Sabale, S. Seader, J. Claiborne Smith, P. Tenenbaum, J. D. Twicken, A. Kamal Uddin, and S. E. Thompson. Measuring Transit Signal Recovery in the Kepler Pipeline. I. Individual Events. *ApJS*, 207(2):35, August 2013.

[63] C. J. Burke, J. L. Christiansen, F. Mullally, S. Seader, D. Huber, J. F. Rowe, J. L. Coughlin, S. E. Thompson, J. Catanzarite, B. D. Clarke, T. D. Morton, D. A. Caldwell, S. T. Bryson, M. R. Haas, N. M. Batalha, J. M. Jenkins, P. Tenenbaum, J. D. Twicken, J. Li, E. Quintana, T. Barclay, C. E. Henze, W. J. Borucki, S. B. Howell, and M. Still. Terrestrial Planet Occurrence Rates for the Kepler GK Dwarf Sample. *ApJ*, 809(1):8, August 2015.

[64] E. A. Petigura, A. W. Howard, and G. W. Marcy. Prevalence of Earth-Size Planets Orbiting Sun-Like Stars. *Proceedings of the National Academy of Science*, 110(48):19273–19278, November 2013.

[65] A. Dattilo, N. M. Batalha, and S. Bryson. A Unified Treatment of Kepler Occurrence to Trace Planet Evolution I: Methodology. arXiv:2308.00103, July 2023.

[66] D. R. Ciardi, D. C. Fabrycky, E. B. Ford, T. N. Gautier III, S. B. Howell, J. J. Lissauer, D. Ragozzine, and J. F. Rowe. On the Relative Sizes of Planets within Kepler Multiple-Candidate Systems. *ApJ*, 763(1):41, January 2013.

[67] L. M. Weiss, G. W. Marcy, E. A. Petigura, B. J. Fulton, A. W. Howard, J. N. Winn, H. T. Isaacson, T. D. Morton, L. A. Hirsch, E. J. Sinukoff, A. Cumming, L. Hebb, and P. A. Cargile. The California-Kepler Survey. V. Peas in a Pod: Planets in a Kepler Multi-planet System Are Similar in Size and Regularly Spaced. *AJ*, 155(1):48, January 2018.

[68] L. M. Weiss and G. W. Marcy. The Mass–Radius Relation for 65 Exoplanets Smaller than 4 Earth Radii. *ApJL*, 783(1):L6, March 2014.

[69] B. J. Fulton, E. A. Petigura, A. W. Howard, H. Isaacson, G. W. Marcy, P. A. Cargile, L. Hebb, L. M. Weiss, J. A. Johnson, T. D. Morton, E. Sinukoff, I.J.M. Crossfield, and L. A. Hirsch. The California-Kepler Survey. III. A Gap in the Radius Distribution of Small Planets. *AJ*, 154(3):109, September 2017.

[70] A. W. Howard, G. W. Marcy, S. T. Bryson, J. M. Jenkins, J. F. Rowe, N. M. Batalha, W. J. Borucki, D. G. Koch, E. W. Dunham, T. N. Gautier III, J. Van Cleve, W. D. Cochran, D. W. Latham, J. J. Lissauer, G. Torres, T. M. Brown, R. L. Gilliland, L. A. Buchhave, D. A. Caldwell, J. Christensen-Dalsgaard, D. Ciardi, F. Fressin, M. R. Haas, S. B. Howell, H. Kjeldsen, S. Seager, L. Rogers, D. D. Sasselov, J. H. Steffen, G. S. Basri, D. Charbonneau, J. Christiansen, B. Clarke, A. Dupree, D. C. Fabrycky, D. A. Fischer, E. B. Ford, J. J. Fortney, J. Tarter, F. R. Girouard, M. J. Holman, J. A. Johnson, T. C. Klaus, P. Machalek, A. V. Moorhead, R. C. Morehead, D. Ragozzine, P. Tenenbaum, J. D. Twicken, S. N. Quinn, H. Isaacson, A. Shporer, P. W. Lucas, L. M. Walkowicz, W. F. Welsh, A. Boss, E. Devore, A. Gould, J. C. Smith, R. L. Morris, A. Prsa, T. D. Morton, M. Still, S. E. Thompson, F. Mullally, M. Endl, and P. J. MacQueen. Planet Occurrence within 0.25 AU of Solar-Type Stars from Kepler. *ApJS*, 201(2):15, August 2012.

[71] C. D. Dressing and D. Charbonneau. The Occurrence of Potentially Habitable Planets Orbiting M Dwarfs Estimated from the Full Kepler Dataset and an Empirical Measurement of the Detection Sensitivity. *ApJ*, 807(1):45, July 2015.

[72] M. Wall. NASA Seeks New Ideas for Ailing Planet-Hunting Spacecraft's Mission. https://www.space.com/22276-nasa-kepler-spacecraft-mission-ideas.html, 2013.

[73] C. Hellier, D. R. Anderson, A. Collier Cameron, A. P. Doyle, A. Fumel, M. Gillon, E. Jehin, M. Lendl, P.F.L. Maxted, F. Pepe, D. Pollacco, D. Queloz, D. Ségransan, B. Smalley, A.M.S. Smith, J. Southworth, A.H.M.J. Triaud, S. Udry, and R. G. West. Seven Transiting Hot Jupiters from WASP-South, Euler and TRAPPIST: WASP-47b, WASP-55b, WASP-61b, WASP-62b, WASP-63b, WASP-66b and WASP-67b. *MNRAS*, 426(1):739–750, October 2012.

[74] J. C. Becker, A. Vanderburg, F. C. Adams, S. A. Rappaport, and H. M. Schwengeler. WASP-47: A Hot Jupiter System with Two Additional Planets Discovered by K2. *ApJL*, 812(2):L18, October 2015.

[75] NASA. NASA Retires Kepler Space Telescope, Passes Planet-Hunting Torch. https://www.nasa.gov/press-release/nasa-retires-kepler-space-telescope-passes-planet-hunting-torch, 2018.

[76] NASA. Kepler Telescope Bids 'Goodnight' with Final Commands. https://www.jpl.nasa.gov/news/kepler-telescope-bids-goodnight-with-final-commands, 2018.

[77] G. R. Ricker, J. N. Winn, R. Vanderspek, D. W. Latham, G. Á. Bakos, J. L. Bean, Z. K. Berta-Thompson, T. M. Brown, L. Buchhave, N. R. Butler, R. P. Butler, W. J. Chaplin, D. Charbonneau, J. Christensen-Dalsgaard, M. Clampin, D. Deming, J. Doty, N. De Lee, C. Dressing, E. W. Dunham, M. Endl, F. Fressin, J. Ge, T. Henning, M. J. Holman, A. W. Howard, S. Ida, J. M. Jenkins, G. Jernigan, J. A. Johnson, L. Kaltenegger, N. Kawai, H. Kjeldsen, G. Laughlin, A. M. Levine, D. Lin, J. J. Lissauer, P. MacQueen, G. Marcy, P. R. McCullough, T. D. Morton, N. Narita, M. Paegert, E. Palle, F. Pepe, J. Pepper, A. Quirrenbach, S. A. Rinehart, D. Sasselov, B. Sato, S. Seager, A. Sozzetti, K. G. Stassun, P. Sullivan, A. Szentgyorgyi, G. Torres, S. Udry, and J. Villasenor. Transiting Exoplanet Survey Satellite (TESS). *Journal of Astronomical Telescopes, Instruments, and Systems*, 1:014003, January 2015.

[78] H. Rauer, C. Catala, C. Aerts, T. Appourchaux, W. Benz, A. Brandeker, J. Christensen-Dalsgaard, M. Deleuil, L. Gizon, M. J. Goupil, M. Güdel, E. Janot-Pacheco, M. Mas-Hesse, I. Pagano, G. Piotto, D. Pollacco, Ċ. Santos, A. Smith, J. C. Suárez, R. Szabó, S. Udry, V. Adibekyan, Y. Alibert, J. M. Almenara, P. Amaro-Seoane, M. Ammler-von Eiff, M. Asplund, E. Antonello, S. Barnes, F. Baudin, K. Belkacem, M. Bergemann, G. Bihain, A. C. Birch, X. Bonfils, I. Boisse, A. S. Bonomo, F. Borsa, I. M. Brandão, E. Brocato, S. Brun, M. Burleigh, R. Burston, J. Cabrera, S. Cassisi, W. Chaplin, S. Charpinet, C. Chiappini, R. P. Church, Sz. Csizmadia, M. Cunha, M. Damasso, M. B. Davies, H. J. Deeg, R. F. Díaz, S. Dreizler, C. Dreyer, P. Eggenberger, D. Ehrenreich, P. Eigmüller, A. Erikson, R. Farmer, S. Feltzing, F. de Oliveira Fialho, P. Figueira, T. Forveille, M. Fridlund, R. A. García, P. Giommi, G. Giuffrida, M. Godolt, J. Gomes da Silva, T. Granzer, J. L. Grenfell, A. Grotsch-Noels, E. Günther, C. A. Haswell, A. P. Hatzes, G. Hébrard, S. Hekker, R. Helled, K. Heng, J. M. Jenkins, A. Johansen, M. L. Khodachenko, K. G. Kislyakova, W. Kley, U. Kolb, N. Krivova, F. Kupka, H. Lammer, A. F. Lanza, Y. Lebreton, D. Magrin, P. Marcos-Arenal, P. M. Marrese, J. P. Marques, J. Martins, S. Mathis, S. Mathur, S. Messina, A. Miglio, J. Montalban, M. Montalto, M.J.P.F.G. Monteiro, H. Moradi, E. Moravveji, C. Mordasini, T. Morel, A. Mortier, V. Nascimbeni, R. P. Nelson, M. B. Nielsen, L. Noack, A. J. Norton, A. Ofir, M. Oshagh, R. M. Ouazzani, P. Pápics, V. C. Parro, P. Petit, B. Plez, E. Poretti, A. Quirrenbach, R. Ragazzoni, G. Raimondo, M. Rainer, D. R. Reese, R. Redmer, S. Reffert, B. Rojas-Ayala, I. W. Roxburgh, S. Salmon, A. Santerne, J. Schneider, J. Schou, S. Schuh, H. Schunker, A. Silva-Valio, R. Silvotti, I. Skillen, I. Snellen, F. Sohl, S. G. Sousa, A. Sozzetti, D. Stello, K. G. Strassmeier, M. Švanda, Gy. M. Szabó, A. Tkachenko, D. Valencia, V. Van Grootel, S. D. Vauclair, P. Ventura, F. W. Wagner, N. A. Walton, J. Weingrill, S. C. Werner, P. J. Wheatley, and K. Zwintz. The PLATO 2.0 Mission. *Experimental Astronomy*, 38(1-2):249–330, November 2014.

[79] NASA. Webb Reveals Steamy Atmosphere of Distant Planet in Exquisite Detail. https://webbtelescope.org/contents/news-releases/2022/news-2022-032, 2022.

[80] W. J. Borucki. Obituary: D. G. Koch (1945–2012). *BAAS*, 54:020, February 2022.

INDEX